AF766540

Forsthoffer's Rotating Equipment Handbooks

Forsthoffer's Rotating Equipment Handbooks

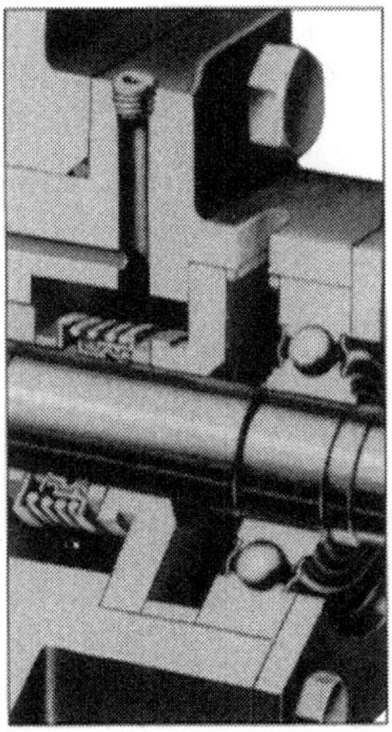

Volume 2: Pumps

W E (Bill) Forsthoffer

President, Forsthoffer Associates
Washington Crossing, PA, USA

ELSEVIER

UK	Elsevier Ltd, The Boulevard, Langford Lane, Kidlington, Oxford OX5 1GB, UK
USA	Elsevier Inc, 360 Park Avenue South, New York, NY 10010-1710, USA
JAPAN	Elsevier Japan, Tsunashima Building Annex, 3-20-12 Yushima, Bunkyo-ku, Tokyo 113, Japan

Copyright © 2005 W E Forsthoffer. Published by Elsevier Ltd. All rights reserved.

ISBN 1-85617-468-9

All rights reserved. No part of this publication may be reproduced, stored in a retrieval system or transmitted in any form or by any means: electronic, electrostatic, magnetic tape, mechanical, photocopying, recording or otherwise, without permission in writing from the publishers.

British Library Cataloguing in Publication Data
A CIP Catalogue record for this book is available from the British Library

No responsibility is assumed by the Author or Publisher for any injury and/or damage to persons or property as a matter of products liability, negligence or otherwise, or from any use or operation of any methods, products, instructions or ideas contained in the material herein.

Published by
Elsevier Advanced Technology,
The Boulevard, Langford Lane, Kidlington, Oxford OX5 1GB, UK
Tel: +44(0) 1865 843000
Fax: +44(0) 1865 843971

Typeset by Land & Unwin Ltd., Northamptonshire

Transferred to digital print 2008

Printed and bound in Great Britain by CPI Antony Rowe, Eastbourne

Contents

Preface

This Series has evolved from my personal experience over the last 40 years with the design, selection, testing, start-up and condition monitoring of Rotating Equipment. Most of the concept figures were originally written on a blackboard or whiteboard during a training session and on a spare piece of paper or I beam during a start-up or a problem solving plant visit.

My entire career has been devoted to this interesting and important field. Then and now more than ever, the cost of rotating equipment downtime can severely limit revenue and profits. A large process unit today can produce daily revenues in excess of US$5 million. And yet, the Operators, Millwrights and Engineers responsible for the safety and reliability of this equipment have not been afforded the opportunity to learn the design basis for this equipment in practical terms. I have also observed in the last ten years or so, that the number of experienced personnel in this field is diminishing rapidly.

Therefore the series objective is to present, in User friendly (easy to access), practical terms (using familiar analogies), the key facts concerning rotating equipment design basis, operation, maintenance, installation and condition monitoring to enable the reader (Engineer, Operator and Millwright) to:

- Understand the effect of process and environmental changes on equipment operation, maintenance and reliability
- Condition Monitor equipment on a component basis to optimize up-time, mean time between failure (MTBF) and mean time to repair (MTTR)
- Select, audit and test the Equipment that will produce highest safety and reliability in the field for the lowest life cycle cost.

The hope is that the knowledge contained in this series will enable Plant Operations, Maintenance and Engineering Personnel to easily access the material that will allow them to present their recommendations to management to solve existing costly problems and produce new projects of optimum reliability.

This volume presents the operation of pumps in a process system (using the concept of pump required and produced head), pump selection for cost effective maximum reliability, eliminating hydraulic disturbances in the design and field operation phases, control and protection, practical component monitoring of performance, bearing, seal and auxiliary system condition to assure optimum pump safety and reliability.

Acknowledgements

This series is a result of interactions with literally thousands of dedicated engineers, machinists, operators, vendors, contractors and students who have been an integral part of my career.

I consider myself very fortunate to have been associated with the best of the best mentors, business associates and dear friends throughout my career. Most especially, in chronological order Dick Salzmann, Bob Aimone, Merle Crane, Walt Neibel, the late Murray Rost, Mike Sweeney and Jimmy Trice. Bob, Merle, Murray and Mike have contributed specifically to the material in this series while Dick, Walt and Jimmy have tactfully kept me on track when necessary.

Special thanks to all of the global machinery vendors who have allowed me to use their material for my training manuals and now this publication.

Last but certainly least; my career would not have been possible without the support, encouragement and assistance from my wife Doris and our children Jennifer, Brian, Eric, Michael and Dara.

A special additional note of thanks to Michael who helped assemble the material, and hopefully learned some in doing so, since he has elected to pursue a career in rotating machinery.

About the author

Bill Forsthoffer began his life-time career in rotating machinery in 1962 with De Laval Turbine Inc. as a summer trainee. After obtaining a Bachelor of Arts degree in Mathematics and Bachelor of Science degree in Mechanical Engineering, during which time he worked for De Laval part time in the Test, Compressor and Steam Turbine Departments, he joined De Laval full time in the Compressor Engineering Department in 1968. He was responsible for all phases of centrifugal compressor component and auxiliary design and also made many site visits to provide field engineering assistance for start up and problem resolution.

Bill joined Mobil Oil Corporate Engineering in 1974 and was responsible for all aspects of rotating equipment specification, technical procurement, design audits, test, field construction, commissioning, start-up and troubleshooting.

After 15 years at Mobil, Bill founded his own consulting company in 1990 and has provided rotating equipment consulting services to over 100 companies. Services include: project reliability assurance, training (over 7,000 people trained) and troubleshooting.

Bill is active in the industry as President of Forsthoffer Associates Inc., frequently writes articles for Turbo Machinery International Magazine and conducts many site specific and public workshops each year.

He can be contacted at bill@forsthofferassociates.com

DEDICATION

TO MENTORS EVERYWHERE –
MAY YOU ENJOY THE RIDE

Introduction

- Introduction
- Objectives
- What does a pump pump?
- What comprises a unit?

Introduction

This book is intended for personnel who are or will be responsible for the engineering application, selection, commissioning, operation, maintenance or reliability for pumps of various types. The importance of pumps in attaining your plant's objectives of maximum unit throughput cannot be overemphasized. Each process unit relies on pumps to meet production quotas. Product revenue losses can range from upwards of a million dollars a day for each day that a process unit is out of operation. In addition, the speeds, pressure, temperatures of fluids associated with pump operation present a potential safety problem. As a result, pump unit safety and reliability is of utmost importance.

Therefore, the objective of this book is to present the principles of pump and auxiliary component design, installation, operation, maintenance and trouble-shooting in a practical manner that will explain why and how pump units react to process and external changes.

Having achieved the book objective, readers will recognize the need to properly select, install, operate and maintain pumps to achieve maximum unit safety, reliability and throughput.

Figure 1.1 Typical pump (Courtesy of Union Pump Co.)

Objectives

Please refer to Figure 1.1. In this figure, a typical pump that may be used in an application in your plant is shown. In many ways, the objectives of this book are similar to what would be done to educate you regarding maintenance or repair of your car. As an example, the more you know about the function of the various parts of your car, the more positive information you will be able to give the mechanic or the better you would be able to repair your car yourself. Therefore, the objectives are as follows:

- To understand the basic principles of pump installation, design and maintenance of each major component.
- To learn how to apply these principles to the safe and reliable operation of pumps on site.

What does a pump pump?

Refer to Figure 1.2 and state what is this pumping? If your answer was a liquid, you are partially correct. As we will see in this book, pumps must pump a fluid in its liquid state. In addition however, every piece of rotating equipment whether it is a pump or compressor that moves a fluid really is moving revenue. That is, Dollars, Ryals, Deutsch Marks, Yen, Rupiahs, etc. Loss of pumping time is loss of product revenue. Therefore reliability and availability of pumps is of utmost importance.

Figure 1.2 Typical pump (Courtesy of Union Pump Co.)

Reliability is the amount of time per year that a pump operates in the process. It is expressed in percent (refer to Figure 1.3).

$$\text{Reliability} = \frac{\text{hours operating time per year}}{\text{8700 hours per year}}$$

Figure 1.3 Reliability

Most users prefer to use Availability rather than Reliability to access the operation of rotating equipment and pumps. The definition of Availability is shown in Figure 1.4.

Availability is the amount of time per year a pump operates divided by the time it is available per year to operate

$$\text{Availability} = \frac{\text{total operating hours per year}}{\text{8700 hours} - \text{planned downtime}}$$

Figure 1.4 Definition of availability

Typical availabilities for pumps are in excess of 95%. These figures will vary significantly depending upon the type of pump and the application.

To give you an idea of how availability can be affected, perform the following simple problem: assume a pump operates at 3,580 revolutions per minute (RPM). Determine how many revolutions this pump will make in one year. Assume there are 8,700 hours per year. The number you obtained is pretty amazing, isn't it? Many millions of revolutions per year. To further appreciate the significance of this requirement on mechanical design, assume the shaft is four inches in diameter so that the circumference of one revolution is approximately one foot. Determine how many miles that pump shaft would roll in one year from the number of revolutions you just calculated. Assume there are 5,280 ft. per mile. For one last exercise, now state how many times that shaft would roll around the world in one year. Assume there are 25,000 miles to transverse the globe once. It is quite amazing isn't it? If you are a maintenance man, now you can feel proud of the services you perform. If you are in operations or engineering consider the mechanical requirement of the equipment. And also consider how important it is to not exceed the capability of the equipment. That is, all equipment is designed with specific limits. Unless we select and operate the equipment within these limits, the normal duty imposed upon this equipment will certainly cause less than desired availability.

What comprises a unit?

Refer to Figure 1.5 and ask yourself – What is G 2301?

Figure 1.5 Typical pump (Courtesy of Union Pump Co.)

If you answered anything but a pump unit, you were not correct. What do we mean by a pump unit? To answer this question perform the following exercise. Take a piece of blank paper and write down all the types of rotating equipment that you can think of. After you have completed this list, then divide that list according to function. That is, which types listed require power, which types provide power, which types transmit power, and which items listed are auxiliary (lubrication, seal systems, cooling systems, etc.). After you have completed this list, you will have four columns listing specific equipment in each column.

Classification	Function
■ Driven	Requires power
■ Driver	Provides power
■ Transmission device	Transmits power
■ Auxiliary system	Provides auxiliary services

Figure 1.6 Functional classification of rotating equipment

Figure 1.6 shows the classifications and functions of rotating equipment. As we can see there are four basic divisions. Now refer back to Figure 1.5 and state what comprises that particular pumping unit. The answer is the pump, the transmission device (coupling), the driver (motor) and the auxiliary systems (pump flush system, lubrication system, cooling water system). The purpose of this exercise was to demonstrate the importance of understanding that pumps are only part of the pumping unit and any troubleshooting exercise must include all four of the items mentioned in this chapter, that is, the driven equipment, the transmission device, the driver and auxiliary systems.

Frequently people overlook the fact that the high maintenance items in pumps which are bearings and seals are not individual items, but are a very small part of the lubrication system and the seal system. Trouble-shooting these components requires that the entire lubrication system and entire seal system be considered. Many times maintenance changes bearings only to find they must do so again within a week. Paying attention to the lubrication system will minimize this action and will assure that all items in the bearing system in this particular instance are considered.

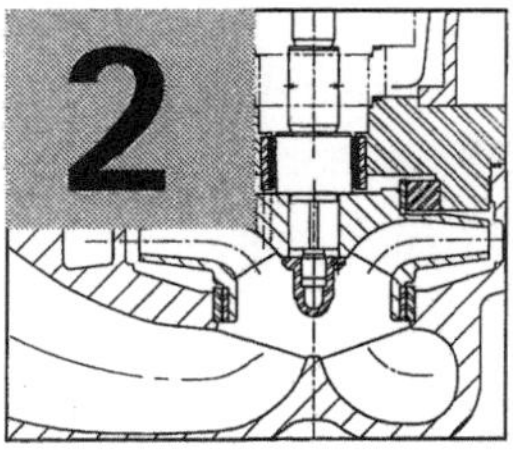

2 Pump types and applications

- Definition of pump types
- Positive displacement pumps
- Dynamic pumps
- When to use positive displacement or centrifugal pumps

Definition of pump types

A pump is defined as a device that moves a liquid by increasing the energy level of the liquid. There are many ways to accomplish this objective. At the conclusion of this chapter one will be able to identify all of the different types of pumps on site and to state the function of each specific type. Refer to Figure 2.1 and observe all of the different types of pumps which will be discussed in this chapter. Note that the pumps are divided into two distinct groups. One group pumps the liquid by means of positive displacement – the other group pumps the liquid by means of dynamic action.

Types of pumps

Positive displacement		Dynamic	
Reciprocating	**Rotary**	**Single stage**	**Multistage**
■ Power	■ Screw	■ Overhung	■ Horizontal split
■ Diaphragm	■ Gear	■ Inline	■ Barrel
■ Metering		■ Integral gear	■ Canned
■ Direct acting		■ Centrifugal	■ Sump
		■ Double flow	■ Submersible
		■ Sump	
		■ Submersible	
		■ Magnetic drive	

Figure 2.1 Types of pumps

Refer to Figure 2.2 for the definition of positive displacement and dynamic action and the characteristics of each type of pump.

Positive displacement – Dynamic pump comparison

	Positive displacement	Dynamic
Definition	Increase pressure by operating on a fixed volume in a confined space	Increases pressure by using rotary blades to increase fluid velocity
Types	Screw, gear, reciprocating	Centrifugal axial
Characteristics	■ Constant volume ■ Variable differential head ■ Relatively insensitive to liquid properties ■ Relatively insensitive to system changes ■ Not self-limiting	■ Variable volume ■ Constant differential head ■ Sensitive to liquid properties ■ Sensitive to system changes ■ Self-limiting
Characteristic flow vs. differential head curves	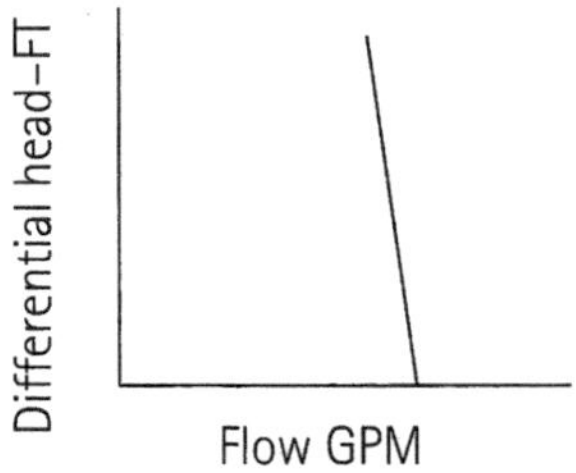	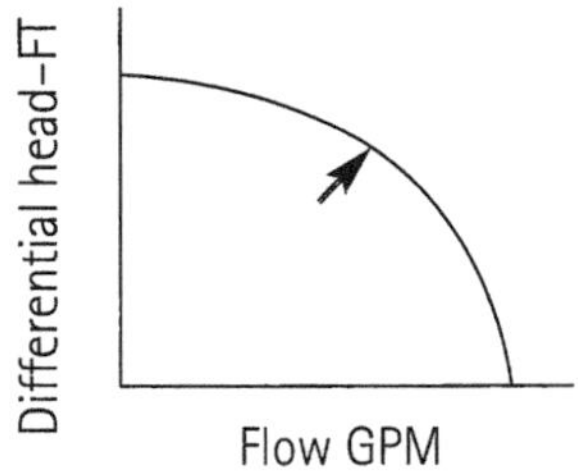

Figure 2.2 Positive displacement – dynamic pump comparison

Regardless of whether the pumps moves the liquid by positive displacement or dynamic means, each pump is divided into a hydraulic and a mechanical end. Figure 2.3 identifies these features for a positive displacement pump.

Figure 2.3 Power pump (Courtesy of Union Pump Co)

One important fact to remember is that the liquid doesn't care how it is moved; that is, the performance relationships of head, horsepower and efficiency will remain the same for all pumps regardless of whether they are positive displacement or dynamic. These relationships will be discussed.

As far as the mechanical end is concerned, the mechanical components of shafts, bearings, seals, couplings and casings perform the same function regardless of the pump type. Figure 2.4 is a chart that shows the similarities between the hydraulic and the mechanical ends of pumps regardless of their type.

Pump similarities – hydraulic and mechanical ends regardless of pump type

Hydraulic end

Head (energy) required

$$HD_{REQ} = \frac{2.31 \times \Delta P}{S.G.}$$

$\Delta P = P_2 - P_1$ (PSIG)

S.G. = specific gravity

$$BHP = \frac{HD \times GPM \times S.G.}{3960 \times \text{pump efficiency}}$$

GPM = U.S. Gallons/minute

S.G. = Specific gravity

Mechanical end

Mechanical components

- Casing (cylinder)
- Seals (mechanical or packing)
- Radial bearings
- Thrust bearing

Figure 2.4 Pump similarities – hydraulic and mechanical ends regardless of pump type

Again it can be seen from this chart that the relationships for pump performance are identical regardless of pump type. The only difference being the efficiency of one pump type relative to another. Secondly, it can be shown that the mechanical ends, housings, bearings, seals, etc. are very similar in each type of pump so that the function of a bearing or seal remains the same regardless of the type of pump.

Positive displacement pumps

As can be seen from Figure 2.2 a positive displacement pump is a constant flow variable head device. Refer to Figure 2.5 which shows a schematic of a double acting piston pump.

As the piston moves from left to right the pressure of the liquid will be increased and the pump will displace the liquid regardless of its specific gravity and viscosity as long as sufficient power is available from the pump driver. The types of positive displacement pumps which can be found in any petrochemical plant, refinery or gas plant are noted below. Refer to Table 2.1 for the application limits of positive displacement pumps.

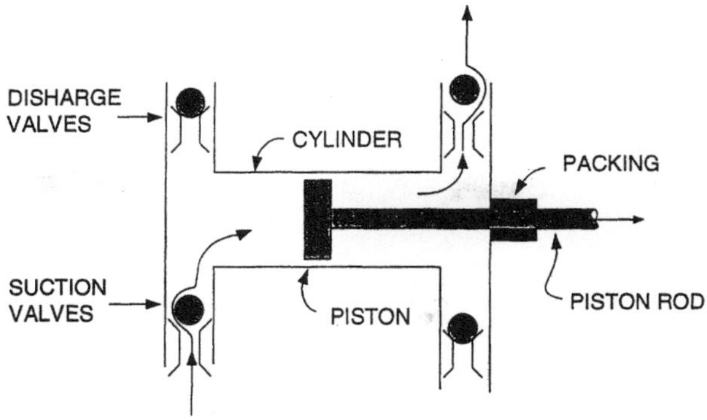

Figure 2.5 Double acting piston pump

Table 2.1 Application envelope positive displacement pumps

Pump type	Pressure-PSIG	Flow rate-GPM	Horsepower (max)
Power pumps			
■ Plunger-(horizontal)	1,000–30,000	10–250	200
■ Plunger (vertical)	1,000–30,000	10–500	1,500
■ Piston (horizontal)	up to 1,000	up to 2,000	2,000
Direct acting			
■ Piston	up to 350	up to 1,000	500
■ Plunger	up to 2,000	50–300	500
Metering	up to 7,500	up to 1,000 gal/hr	10*
Rotary			
■ Gear	up to 200	up to 300	50 HP
■ Screw	up to 5,000	up to 5,000	1,000 HP

* horsepower per cylinder, some applications use multiple cylinders

Reciprocating pumps

Reciprocating pumps are those types of positive displacement pumps that increase liquid energy by a pulsating action. The types are power pumps, direct acting steam pumps, diaphragm pumps and metering pumps. All reciprocating pumps produce pulsations that can cause damage to the pumps and/or process system if the system is not properly analyzed and designed. Anti pulsation devices (volume bottles, orifices or pulsation bottles) are usually required.

Figure 2.6 Power pump (Courtesy of Union Pump Co)

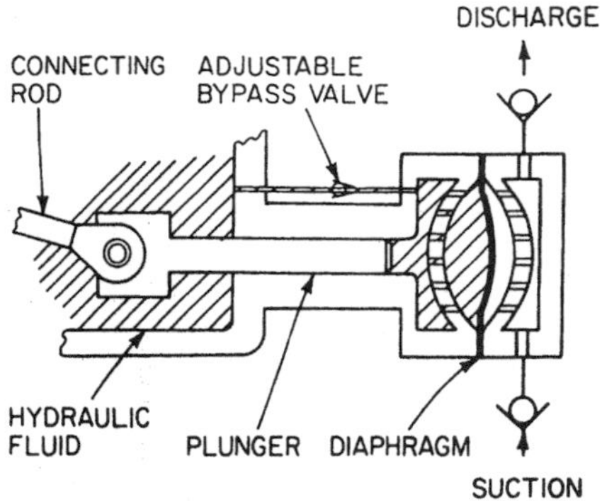

Figure 2.7 Diaphragm pump

Power pumps

A picture of a power pump is shown in Figure 2.6. Power pumps are used normally for high pressure, low flow applications, typically carbonate, amine service or high pressure water or oil services. They can either be horizontal or vertical. The major parts of a power pump as shown in Figure 2.6 include the liquid cylinder with pistons and rods, the valves and power end. The power end consists of the crankshaft with bearings, connecting rod and crosshead assembly. It is termed, 'Power Pump' because it is driven by an external power source, such as an electric motor, or internal combustion engine, instead of steam cylinders as in direct-acting pumps.

Diaphragm pumps

A schematic of a diaphragm pump is shown in Figure 2.7. In this type of pump the power end and liquid end areas are approximately the same. This results in the pump being capable of pumping against pressures no greater than that of the motive fluid. This pump has limited use in the refining and petrochemical industry and is used primarily for metering services.

Metering pumps

A diaphragm type metering pump or 'proportioning' pump is shown in Figure 2.8.

This type of pump is most commonly used for chemical injection service when it is required to precisely control the amount of chemical or inhibitor being injected into a flowing process stream. Volume control is provided by varying the effective stroke length. There are two basic types of metering pumps:

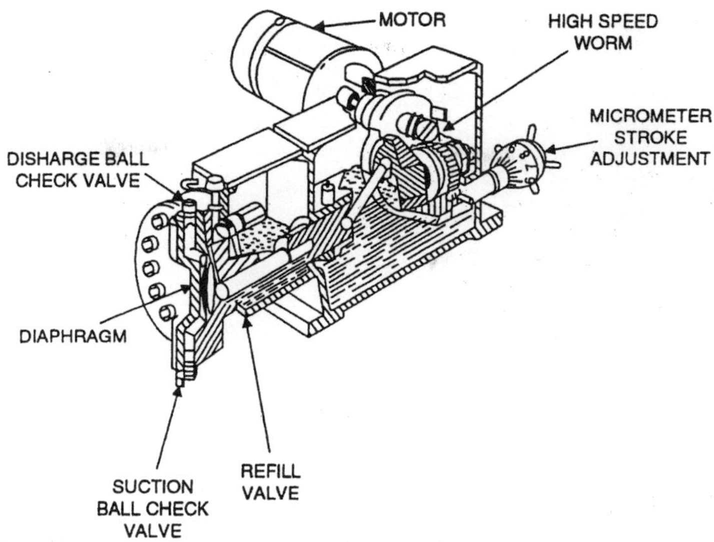

Figure 2.8 Diaphragm type of metering pump

1 Packed plunger pump – the process fluid is in direct contact with the plunger and is used for higher flow applications.

2. Diaphragm pump – process fluid is isolated from the plunger by means of a hydraulically actuated flat or shaped diaphragm and is used for lower flow applications or where escape of the pumped liquid to atmosphere is not acceptable.

Metering pumps can be furnished with either single or multiple pumping elements.

When the pumped liquid is toxic or flammable, diaphragm pumps can be provided with double diaphragms with a leak detector to alarm on failure of either diaphragm. The American Petroleum Institute has published standard 675 which covers the minimum requirements for controlled volume pumps for use in refinery service.

Rotary pumps

There are a number of different types of pumps which are classified as 'rotaries'. Rotary pumps are positive displacement pumps that do not cause pulsation. The inherent high efficiency and versatility of the 'rotary' (screw, gear and others) makes this design very suitable for use in lube oil, seal oil and other high viscosity oil services. They can handle capacities from a fraction of a gallon to more than 5,000 GPM, with pressures ranging up through 5,000 PSI when pumping liquids with viscosities from les than one (1) centistoke to more than 1,000,000 SSU.

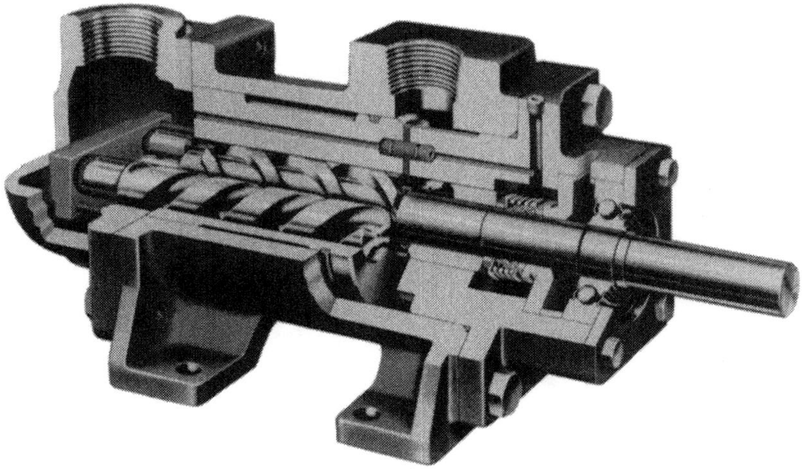

Figure 2.9 Screw pump (Courtesy of IMO Industries)

Screw pumps

Figure 2.9 illustrates the screw pump design. Fluid flow is carried axially between the threads of two or more close clearance rotors so that a fixed volume of fluid is displaced with each revolution. This design is frequently used for lube and seal service.

Gear pumps

A picture of a commonly used gear pump is shown in Figure 2.10. With this type of pump, fluid is carried between the teeth of two external gears and displaced as they mesh. Gear pumps are used for small volume lube oil services and liquids of very high viscosity (asphalt, polyethylene, etc).

Figure 2.10 Gear pump (Courtesy of IMO Industries)

Dynamic pumps

Centrifugal pumps can be referred to as 'dynamic' machines. That is to say they use centrifugal force for pumping liquids from one level of pressure to a higher level of pressure. Liquid enters the center of the rotating impeller, which imparts energy to the liquid. Centrifugal force then discharges the liquid through a volute as shown in Figure 2.11.

The centrifugal pump is one of the most widely used fluid handling devices in the refining and petrochemical industry. Every plant has a multitude of these types of pumps operating. A brief description of the various designs found in operating plants follows. Refer to Table 2.2 for the application limits of dynamic pumps.

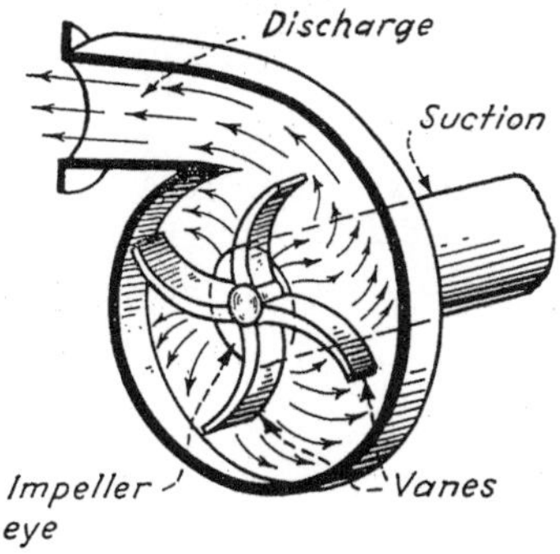

Figure 2.11 Dynamic pump principle

Table 2.2 Application limits – Dynamic pumps

Pump type	Pressure PSIA	Head (FT)	Flowrate-GPM	Horsepower-BHP
■ Single stage overhung	300	800	7,000	2,000
■ Single stage double Flow between bearing	300	800	>70,000	>8,000
■ Single stage inline	300	800	7,000	200
■ Integral gear centrifugal	1,000	2,500	2,000	400
■ Multistage horizontal Split	2,000	6,000	3,000	500
■ Multistage barrel	3,000	8,000	2,000	>5,000
■ Vertical canned pump	1,500	4,000	>70,000	1,000
■ Sump pumps	100	300	7,000	250
■ Submersible	100	300	4,000	150
■ Magnetic drive pump	300	800	3,000	800

Single stage overhung pump

The single stage overhung pump design shown in Figure 2.12 is probably the most widely used in the industry. Its construction incorporates an impeller affixed to a shaft which has its center of gravity located outside the bearing support system.

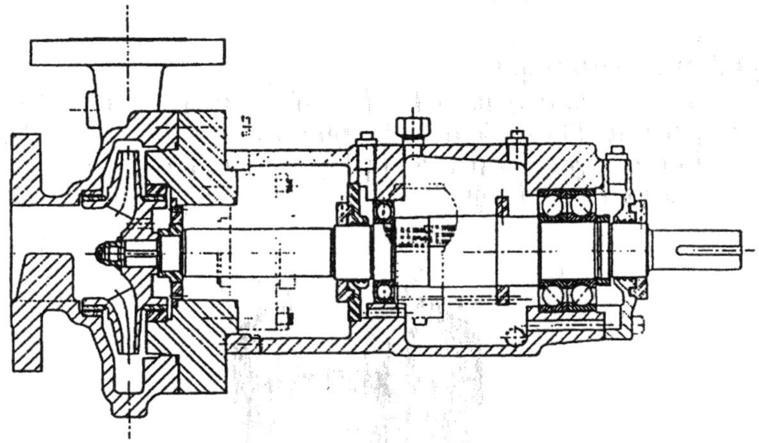

Figure 2.12 Single stage overhung pump (Courtesy of Union Pump Co)

Single stage inline

This type of pump is finding increased usage in applications of low head, flow and horsepower. Refer to Figure 2.13.

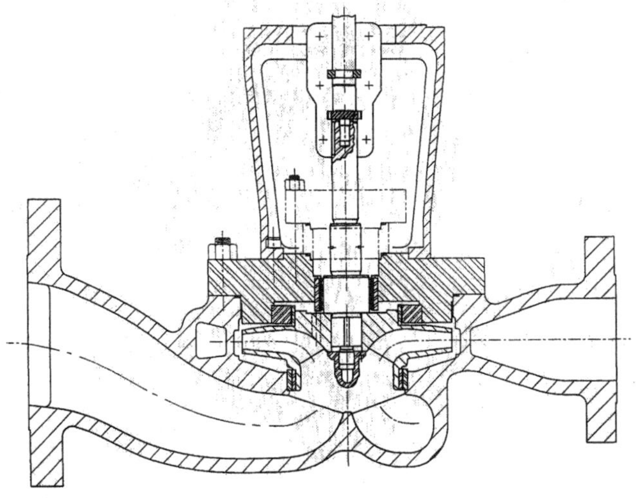

Figure 2.13 Single stage inline (Courtesy of Union Pump Co)

The advantage of this pump design is that it can be mounted vertically (inline) between pipe flanges and does not require a baseplate. A concrete, grouted support plate however, is strongly recommended. It should be noted that many inline designs do not incorporate bearings in the pump and rely on a rigid coupling to maintain pump and motor shaft alignment. Acceptable pump shaft assembled runout with these types of pumps should be limited to 0.001".

Integral gear centrifugal

This type of pump is used for low flow applications requiring high head. Refer to Figure 2.14. The pump case design is similar to the inline, but incorporates pump bearings and an integral gear to increase impeller speeds over 30,000.

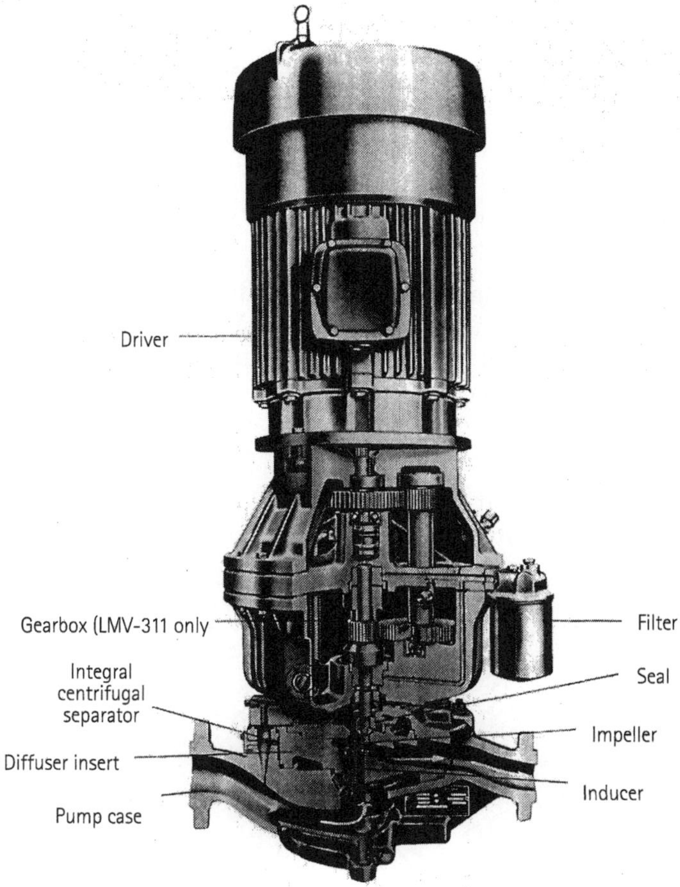

Figure 2.14 Integral gear centrifugal pump (Courtesy of Sunstrand Corp.)

Single stage double flow between bearing

As the name implies, double suction impellers are mounted on between-bearing rotors as shown in Figure 2.15.

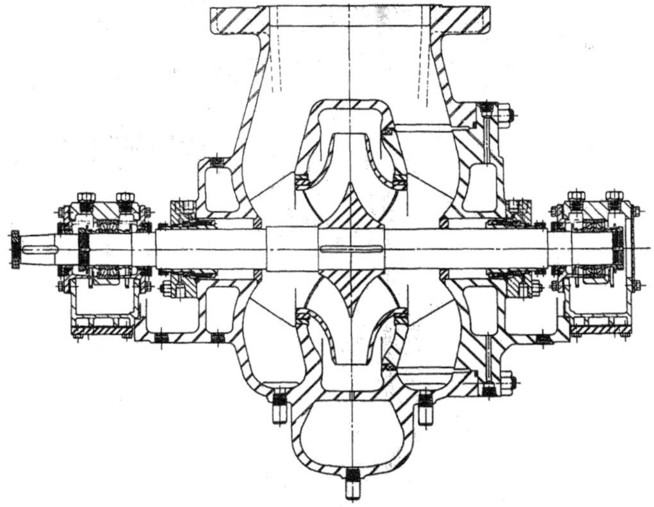

Figure 2.15 Single stage double flow between bearing (Courtesy of Union Pump Co)

This pump design is commonly used when flow and head requirements make it necessary to yield low values of NPSH required. When designing piping systems for this type of pump, care must be taken to assure equal flow distribution to each end of the impeller to prevent cavitation and vibration.

Multistage horizontal split

When the hydraulic limits of a single stage pump are exceeded, it is common practice to use a multistage pump shown in Figure 2.16.

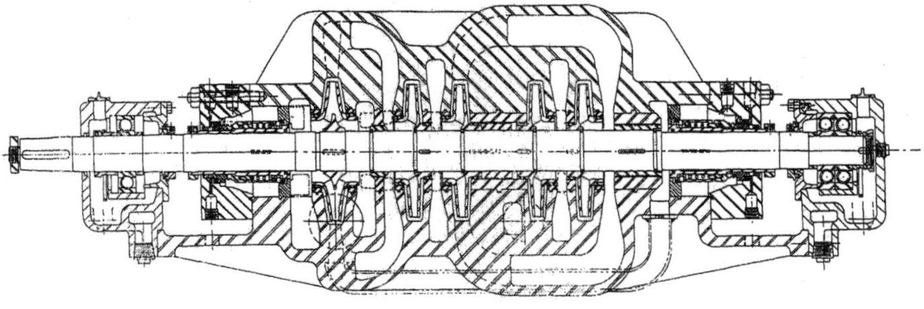

Figure 2.16 Multistage centifugal pump (Courtesy of Union Pump Co)

This figure illustrates a horizontal split casing design which allows the rotor to be removed vertically after the top half casing is unbolted. This type of pump is normally limited to working pressure of approximately 2000 PSI, temperatures to 600°F and S.G. of 0.7 or greater. Impeller configuration for this type of pump can be either 'inline' or 'opposed'. The 'opposed' impeller arrangement has the advantage of not requiring a thrust balancing device which is required for the 'inline' configuration.

Multistage barrel

The so called barrel casing design is shown in Figure 2.17.

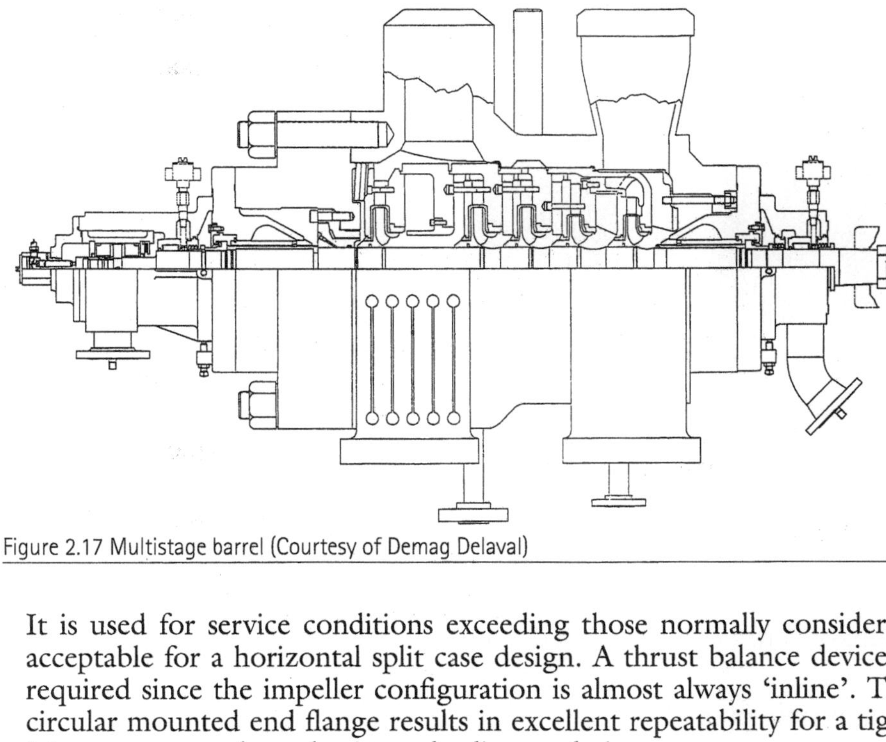

Figure 2.17 Multistage barrel (Courtesy of Demag Delaval)

It is used for service conditions exceeding those normally considered acceptable for a horizontal split case design. A thrust balance device is required since the impeller configuration is almost always 'inline'. The circular mounted end flange results in excellent repeatability for a tight joint as compared to a horizontal split case design.

Vertical canned pump

There are a variety of vertical pump types and designs. Use of the vertical pump is usually dictated by low available NPSH. To provide adequate NPSH characteristics, the first stage impeller can be lowered. The can-type design shown in Figure 2.18 is one that is widely used in the Petrochemical industry, particularly when pumping low specific

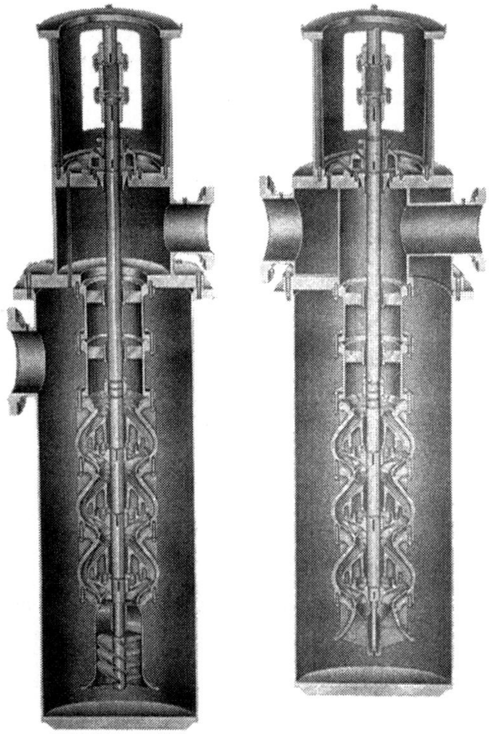

LEFT Figure 2.18 Vertical multistage pump (Courtesy of Dresser-Pacific Pumps)

gravity liquids from tank farm facilities. The multistage can pump is comprised of a number of bowl assemblies all contained within a can. This reduces the risk of hydrocarbon leakage to atmosphere. Lubrication to the sleeve bearings located throughout the length of the shaft is provided by the pumped liquid. Therefore, bearing material vs product compatibility must always be considered. It is necessary however to prevent leakage where the shaft passes through the can to connect to the driver. Sealing is normally provided using a mechanical seal.

Sump pump

The sump pump illustrated in Figure 2.19 is a popular design for handling run off streams of rain water, non corrosive or corrosive liquids. The setting limitation for this cantilever design is approximately 10 ft. This particular design incorporates an enclosed lineshaft with external lubrication to the bottom bearing. The pump shaft and impeller are coupled to the driver which is supported by a motor support bracket bolted to a cover plate.

Submersible pumps

This type of pump consists of an electric motor driver is coupled directly to the impeller/bowl assembly (see Figure 2.20). All components are designed to be submerged in the pumped fluid. In the past, this type of pump did not find widespread use in the refining and petrochemical industry.

However, with increasing environmental restrictions, this type of pump is being used more frequently in the refining and petrochemical industry.

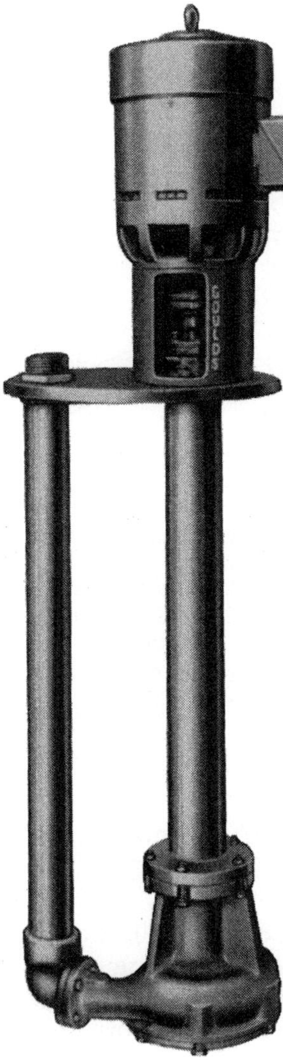

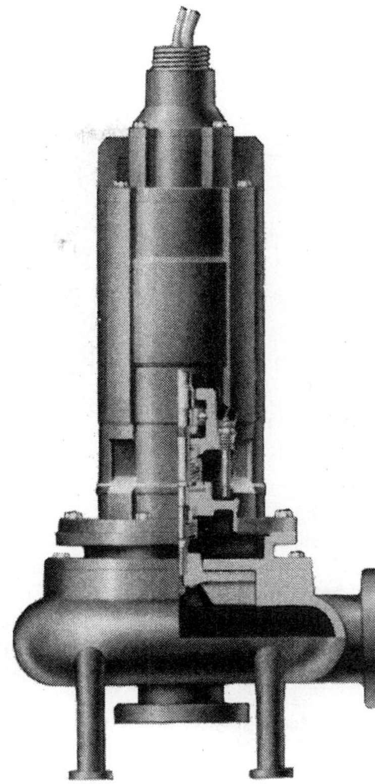

ABOVE Figure 2.20 Submersible Pump
(Courtesy of Goulds pumps)

LEFT Figure 2.19 Cantilever design sump pump
(Courtesy of Goulds pumps)

Magnetic drive pumps

As a result of more stringent environmental constraints and regulations, sealless pump technology has gained prominence. One such design is the magnetic drive pump (MDP), shown in Figure 2.21. This is a design such that the motor shaft is attached to the power frame of the magnetic drive pump by means of a flexible or rigid coupling. The outer magnet and shaft assembly is supported by its own bearings. Alignment requirements for this type pump are similar to that for horizontal mounted centrifugal pumps fitted with mechanical seals or packing. The sealless pump is generally applied when there is a need to contain toxic or hazardous fluids.

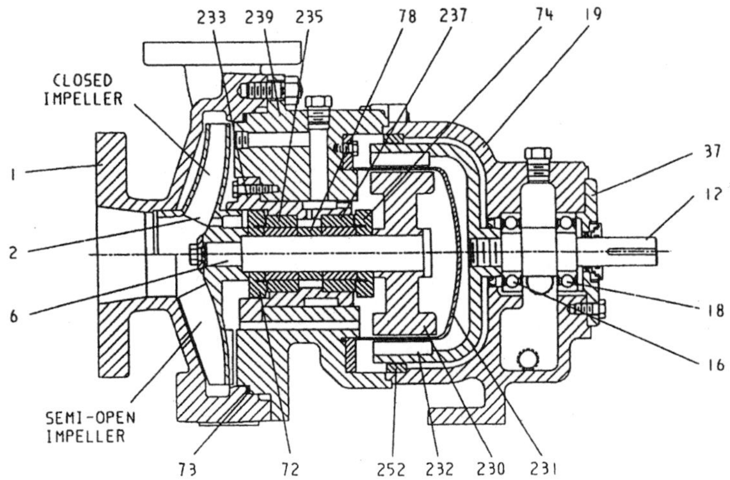

1 Casing	37 Collar, bearing outboard	232 Magnet assembly, outer
2 Impeller	72 Collar, thrust inboard	233 Housing, bearing
6 Shaft, pump	73 Gasket	235 Bushing, bearing inboard
12 Shaft, drive	74 Collar, thrust outboard	237 Bushing, bearing outboard
16 Bearing, inboard	78 Spacer, bearing	239 Cover, casing
18 Bearing, outboard	230 Magnet assembly, inner	251 Ring, rub
19 Frame	231 Shell containment	

Figure 2.21 Magnetic drive pump (Courtesy of Hydraulic Institute)

When to use positive displacement or centrifugal pumps

Coverage chart

Selecting the class of pump to use for a specific application can be somewhat confusing to engineers at times. When deciding on the class of pump to use, consideration must be given to hydraulic conditions (head, capacity, viscosity) and the fluid to be handled (corrosive, non-corrosive). The chart in Figure 2.22 provides a guideline for the range of coverage for different classes of pumps.

Obviously, there are overlapping areas and one must be familiar with the hydraulic and mechanical characteristics of each pump to properly apply it to the system. It can be seen that rotary and centrifugal pumps overlap in capacity range of 5,000 GPM and pressure range of 500 PSI. Centrifugal pumps can handle liquid viscosities up to 3,000 SSU, while rotary type pumps normally move materials from 60 SSU to millions of SSU.

PD pumps vs centrifugal pumps – advantages/disadvantages of each

The question of why use a PD pump in place of a centrifugal or vice

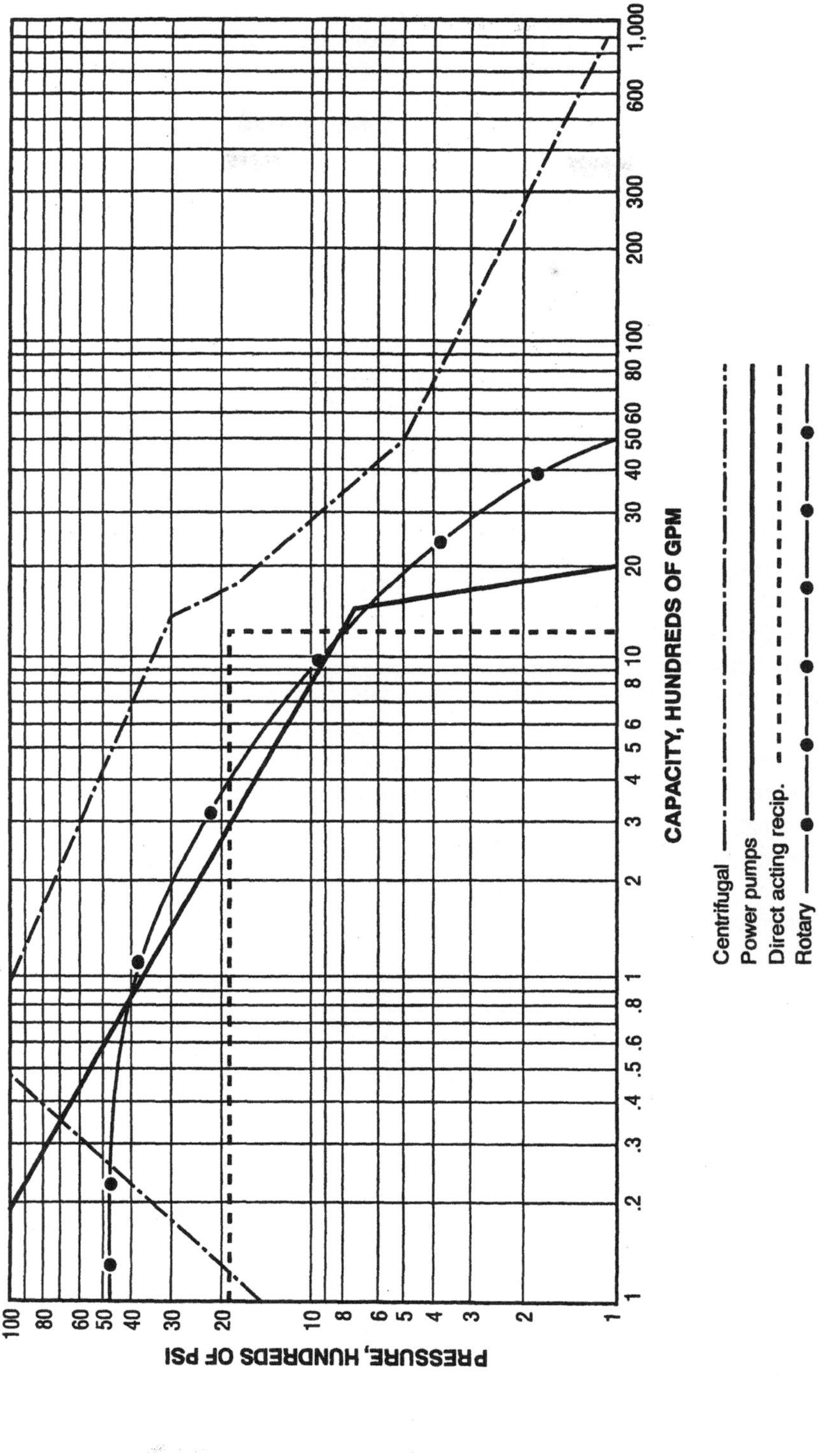

Figure 2.22 Operating range of typical pump types

versa, is often raised. Table 2.3 identifies some advantages and disadvantages associated with each class or equipment which may help make the decision less difficult.

Table 2.3 Advantages/disadvantages of PD pumps vs centrifugal pumps

Advantages	Disadvantages
Centrifugal pumps	
■ Variable capacity control over operating range at constant speed ■ Can handle liquids containing catalyst, dirt solids ■ Can pump liquids with poor lubricity ■ Weight, size, initial cost and installed cost is lower than PD pump with same hydraulic conditions ■ Liberal clearances, no rubbing parts, minimum wear – higher availability ■ Does not normally require overpressure protection over operating range at constant speed	■ Flow rate is effected by specific gravity ■ Viscosity affects performance ■ Requires priming ■ Develops limited head over operating range at constant speed ■ Low to moderate efficiencies
Positive displacement	
■ Not limited to delivery pressure for given capacity ■ Can handle high viscosity liquids efficiently ■ Higher efficiencies than centrifugal ■ Flow rate is not significantly affected by specific gravity	■ Requires over-pressure protection ■ Flow controls with bypass or speed ■ Pulsations associated with reciprocating PD pumps

3 Pump characteristics – positive displacement vs centrifugal (kinetic)

- Introduction
- Positive displacement pumps
- Centrifugal (kinetic) pumps

Introduction

In this chapter we will discuss characteristics of positive displacement pumps and centrifugal (kinetic) pumps. In addition, we will cover the effects of viscosity, specific gravity and system resistance on performance.

Positive displacement pumps

Definition

Since the positive displacement pump is a constant flow, variable head device, its performance characteristic can be represented essentially by a straight line (excluding losses or slip) with variations in differential pressure across the pump. Flow from a PD pump can be produced either through rotary or reciprocating action. Rotary action discharges smooth flow, whereas reciprocating pumps produce pulsating flow characteristics. Adding multiple cylinders and timing the pumping

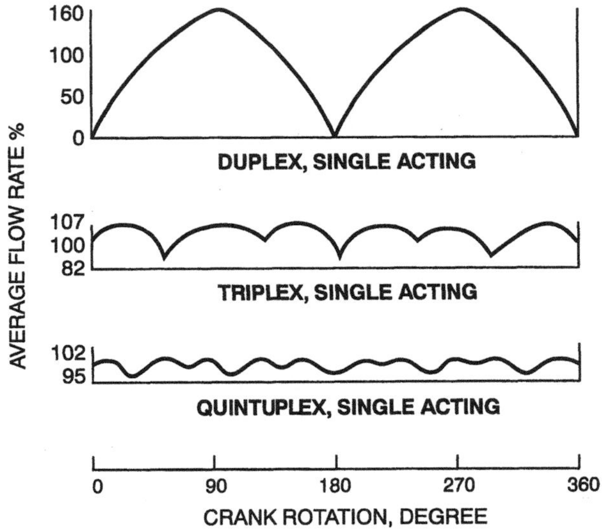

Figure 3.1 Flow characteristics of single acting duplex and triplex power pump

strokes for small overlaps, will significantly reduce the pulsations in reciprocating pumps (refer to Figure 3.1).

Development of characteristics

The actual flow from a PD pump is less than the theoretical quantity by the amount of losses (slip) associated with the specific type of PD pump. The losses associated with reciprocating pumps include valve and seal (rod, piston or plunger) leakage. Leakage areas in a rotary type pump depend on the specific pump design. Slippage in an external gear pump occurs in three areas; between the sides of the gears and sideplate, between the gears and the bodies and between the gears themselves. Slip (losses) in a rotary screw type pump design is a result of the internal clearance between the rotors and the housing; lower viscosities and higher pressures increase slip, which reduces the delivered capacity at a given speed (refer to Figures 3.2, 3.3 and 3.4).

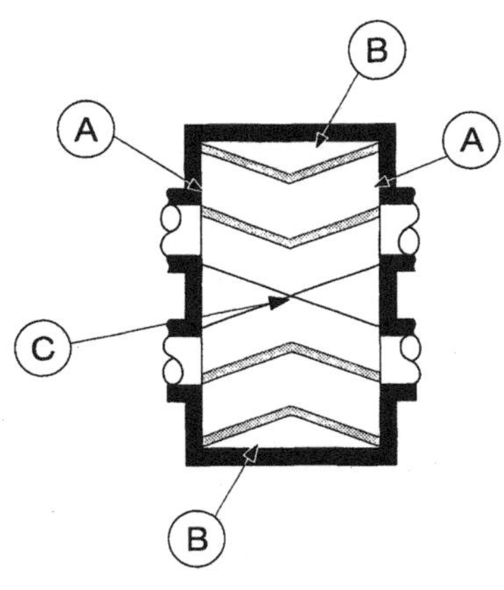

Figure 3.2 Areas of slip in gear pump

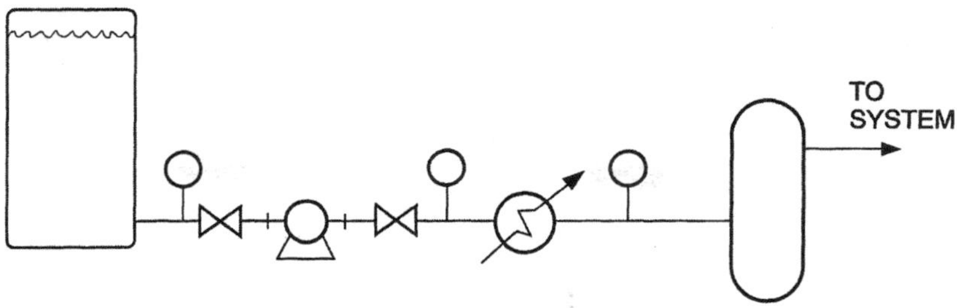

Figure 3.3 Simple system

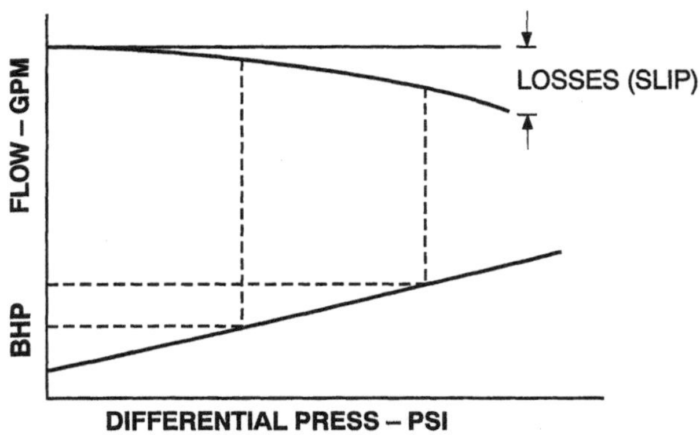

Figure 3.4 How slip affects PD pump performance

The effect of liquid characteristics on performance

Viscosity does impact on the performance characteristics of positive displacement pumps. However, the effect is not as great as for kinetic pumps. As the viscosity of a fluid, handled by a reciprocating pump is increased, valve action is affected. Higher valve leakage occurs at constant pump speed resulting in lower volumetric efficiency. To maintain the same volumetric efficiency, pump speed will have to be reduced as viscosity increases. An additional effect on a reciprocating pump is increased head loss through the suction valves and port, thus increasing the NPSH required.

Rotary pump performance is also affected by changes in fluid viscosity. Capacity is reduced by decreasing viscosity. This occurs because of increased leakage through internal clearance, resulting in a higher slip factor. As viscosity is increased the capacity curve approaches the theoretical displacement.

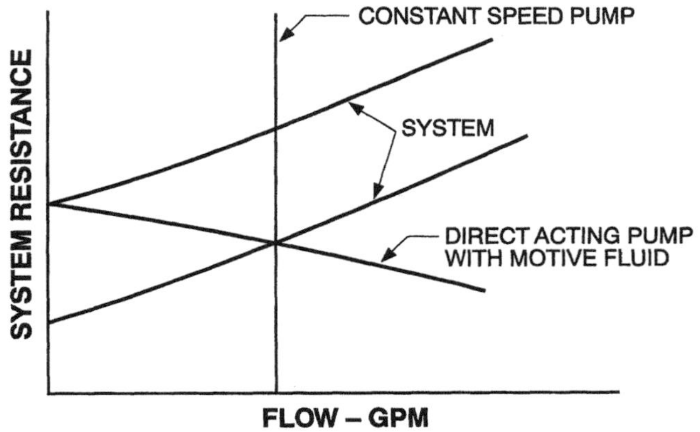

Figure 3.5 Positive displacement pumps. System resistance vs pump performance

The effect of system resistance on performance

Regardless of the type of positive displacement pump, changes in specific gravity do not affect performance. Since the energy produced by a positive displacement pump is unlimited, a positive displacement pump will always meet pressure requirements regardless of the liquid specific gravity. This assumes sufficient power is supplied to drive the pumps. PD pumps deliver essentially constant flow regardless of system friction when handling a given fluid at constant pump speed. The exception is a direct acting pump with motive fluid (steam pump) at fixed pressure. A flow reduction occurs with increasing resistance until a point is reached where the pump stalls. Figure 3.5 illustrates the system interaction with both types of pumps. Figure 3.6 presents a summary of PD pump characteristics and Figure 3.7 presents PD pump performance parameters and Figure 3.7A illustrates the calculation method for determining stall pressure.

Summary of PD pump characteristics

- Constant flow
- Variable head
- Not self limiting
- Pulsating flow (non-rotary types)
- Throughput does not vary with SG change

Figure 3.6 Summary of PD pump characteristics

Summary of PD pump performance parameters

Displacement – slip = capacity

$$\text{Volumetric efficiency} = \frac{\text{capacity} \times 100}{\text{displacement}}$$

Brake hp = friction hp + hydraulic hp

$$\text{Mechanical efficiency} = \frac{\text{hydraulic hp}}{\text{brake hp}}$$

$$= \frac{\text{Gpm} \times \Delta\text{psi}}{1714 \times \text{bhp}}$$

$$= \frac{\text{gpm} \times \text{head (ft)} \times \text{sg}}{3960 \times \text{bhp}}$$

Figure 3.7 Summary of PD pump performance parameters

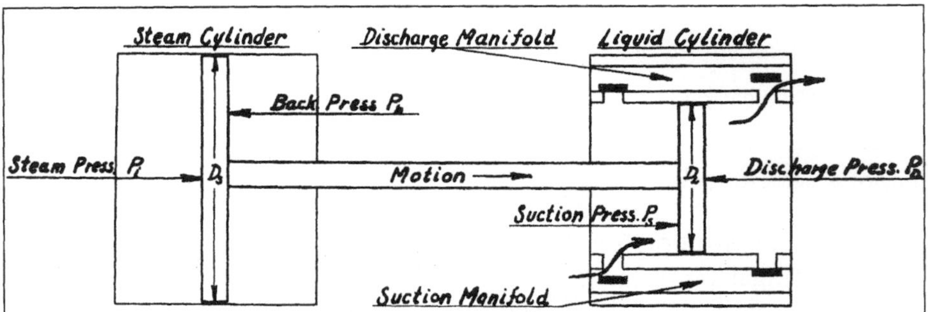

$(P_I - P_b) D_S^2 \times E_M = (P_D - P_S) D_L^2$ — BALANCE OF PRESSURES[1]

$$\text{NET STEAM PRESSURE} = \frac{\text{LIQUID PISTON AREA} \times \text{NET DISCHARGE PRESSURE}}{\text{STEAM PISTON AREA} \times \text{MECHANICAL EFFICIENCY}}$$

$$\text{OR } (P_I - P_b) = \frac{D_L^2}{D_S^2} \times \frac{(P_D - P_S)}{E_M}$$

$$\text{MAX. } P_D \text{ OR STALLING PRESSURE} = (P_I - P_b) \times \frac{D_S^2}{D_L^2} + P_S$$

WHERE D = DIAMETER, IN INCHES
E_M = MECHANICAL EFFICIENCY

(1)–THE RATE OF MOTION OF THE PUMP IS PROPORTIONAL TO THE UNBALANCE OF THE FORCES ACTING ON THEM.

Figure 3.7A Schematic diagram of direct-acting steam pump (Courtesy of Union Pump Co.)

Centrifugal (kinetic) pumps

Definition

The next concept to be covered is that of centrifugal (kinetic) pumps. Refer to Figure 3.8 for the definition and different types of kinetic pumps.

Kinetic (centrifugal) pumps

Delivers a variable amount of liquid for a fixed head (differential pressure) at a specific flow condition by using rotating blades or vanes to increase the velocity of the liquid.

Types include:

- Centrifugal
  - Horizontal (overhung, double suction, multistage)
  - Vertical (inline, integral, sump, can, turbine)
  - Axial
- Regenerative
- Rotating casing (pitot tube type)

Figure 3.8 Kinetic (centrifugal) pumps

Development of pump characteristics

A centrifugal pump adds energy and moves liquid by action of high speed rotating vanes. It delivers a variable amount of liquid for a fixed head (differential pressure). Actually, there are two components to the motion imparted to the liquid by the vanes. One motion is radial in direction outward from the center of the impeller. Also, as the liquid leaves the impeller, it tends to move in a direction tangential to the outside diameter of the impeller. The actual direction which the liquid takes is a result of the two flow directions. Figure 3.9 illustrates this concept.

We can say that the amount of energy added to the liquid by the rotating impeller is related to the velocity with which the liquid moves. We also find that the energy expressed as pressure energy will be proportional to the square of the resultant exit velocity as follows:

$$H \propto \frac{V^2}{2g}$$

where: H = Energy in $\frac{Ftlb_f}{Lb_m}$ of liquid

V = Velocity in Ft. per sec.

g = Acceleration due to gravity in FT/Sec^2

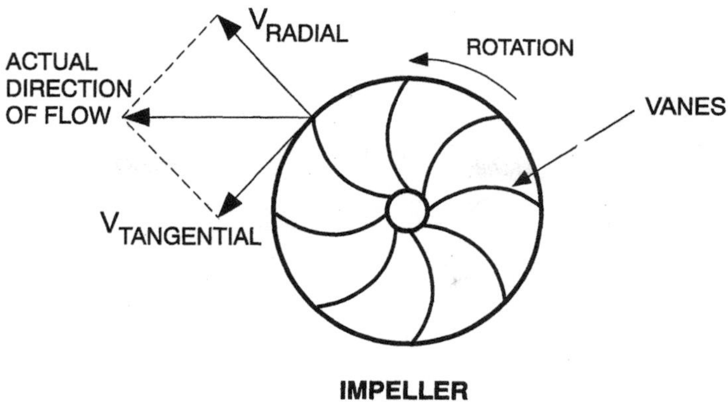

Figure 3.9 Pump impeller discharge velocities

Characteristics

Each centrifugal pump has, for a particular speed and impeller diameter, fluid handled and viscosity, a rating curve which defines the relationship between the head (energy) produced by the pump and the flow through the pump. The total head (energy) the pump is capable of producing is increased as the flow is reduced. A complete characteristic curve also includes efficiency, brake horsepower and $NPSH_r$ (net positive suction head required) curves (refer to Figures 3.10 and 3.11).

The performance curve can be classified as stable or unstable. A stable curve is one in which the head is continuously rising from rated point capacity back to shutoff. An unstable curve is one where the head at shutoff is lower than the maximum developed head. This is an important characteristic for engineers to be familiar with and to avoid if

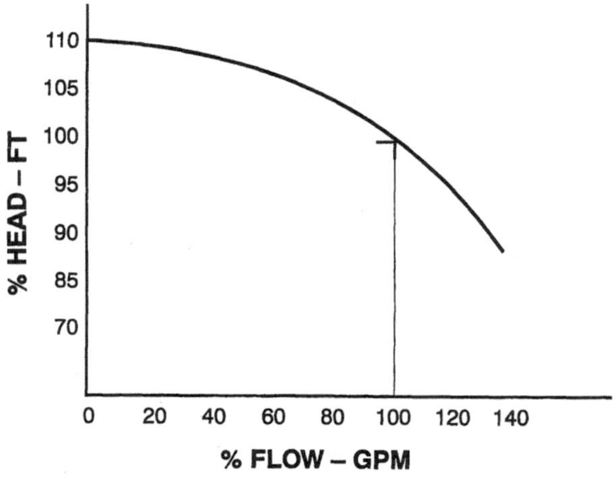

Figure 3.10 Pump characteristics performance curve – centrifugal (typical)

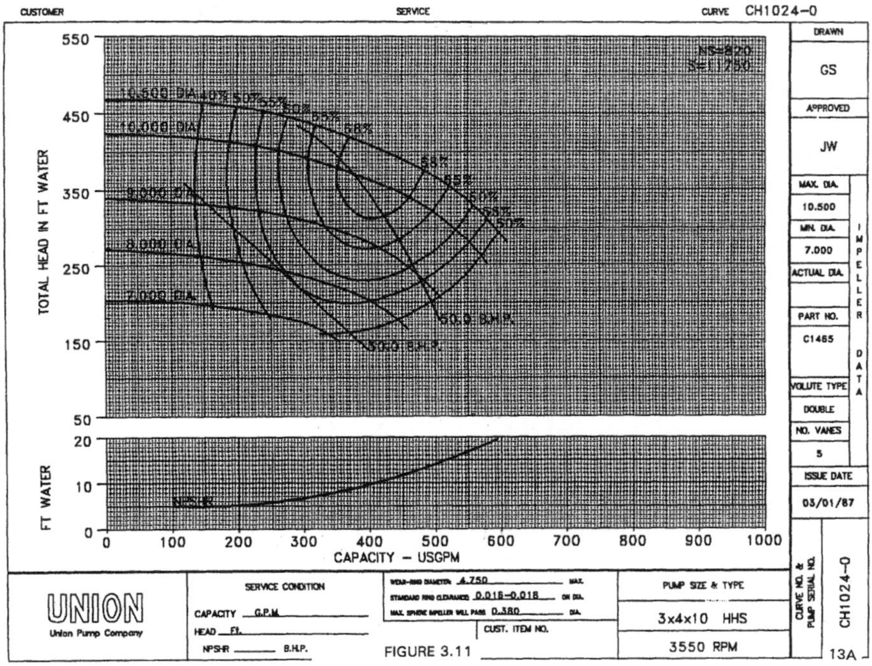

Figure 3.11 Pump characteristics performance curve – centrifugal (typical) (Courtesy of Union Pump Co.)

possible. Operation in the unstable area may result in head capacity fluctuations or potential failure of the pump (refer to Figure 3.12).

The effect of liquid characteristics on performance

The effect of increased viscosity on centrifugal pump performance can be significant. Higher viscosity increases the friction losses, hence the

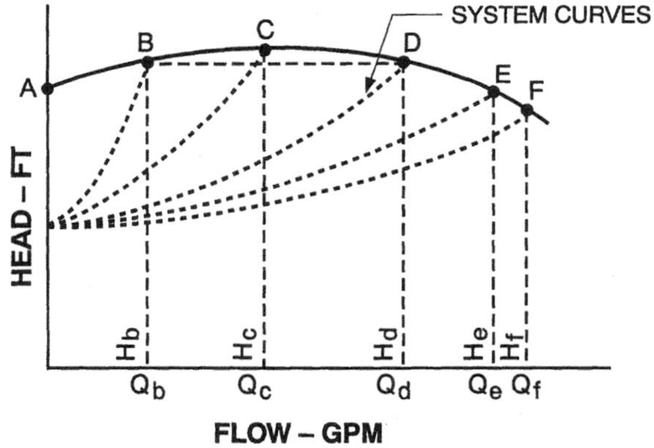

Figure 3.12 Pump characteristics unstable performance

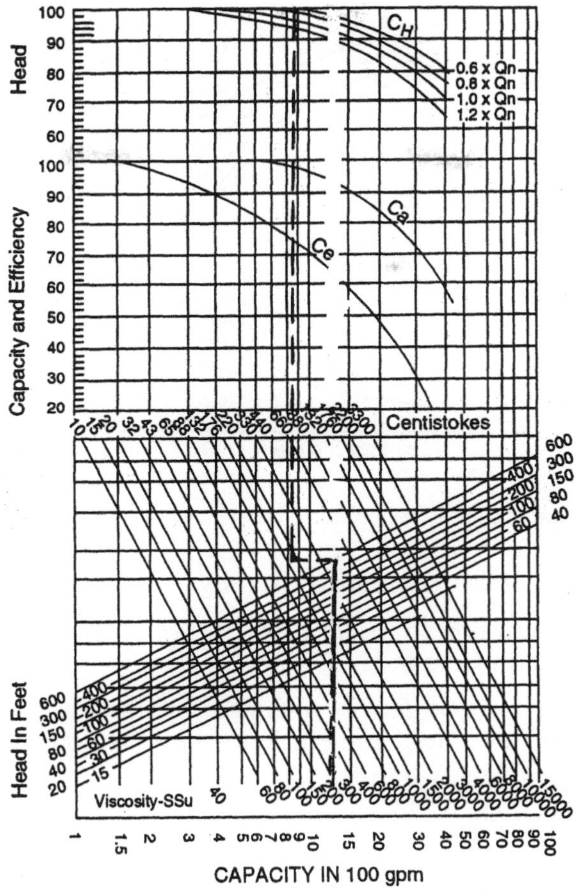

Figure 3.13 Viscosity correction factors (Courtesy of The Hydraulics Institute)

power. In addition, the head capacity curve drops off as viscosity is increased. Based on higher power, and less head, the overall effect is a significant drop in efficiency. The Hydraulics Institute has published curves for estimating pump performance with liquids more viscous than water. Refer to Figure 3.13.

There are two (2) application methods for using the viscosity correction nomograph, they are:

- Initial pump selection
- Determining viscose performance characteristics for an existing pump.

The procedure to use in pump selection is as follows:

1. Locate pump flow – G.P.M.
2. Proceed vertically to pump head – Ft.

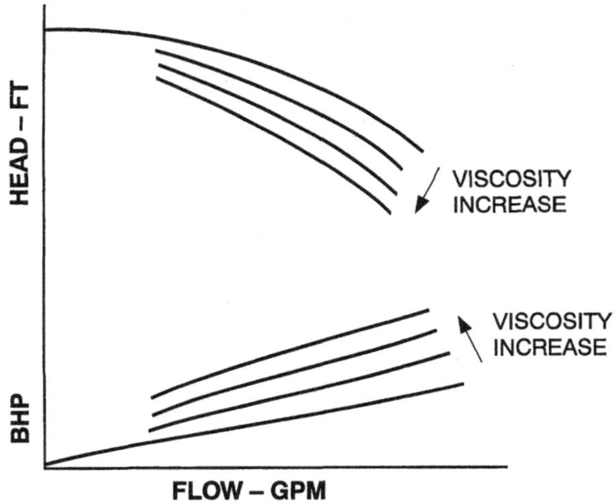

Figure 3.14 Pump characteristics viscosity effects on performance

3. Move horizontally to specified viscosity – SSU or CTS.
4. Draw a vertical line to locate Efficiency (Ce), Flow (Ca) and Head (Ch) correction factors.
5. Using the correction factors obtained for flows from 60% to 120% of flow, generate a head and efficiency curve vs flow for the specified viscosity.

The procedure to use for an existing pump is as follows:

1. Locate pump flow – G.P.M.
2. Proceed vertically to pump head – Ft.
3. Move horizontally to specified viscosity – SSU or CTS.
4. Draw a vertical line to locate Efficiency (Ce), Flow (Ca) and Head (Ch) correction factors.
5. Multiply values on the pump curve (tested on water) by the appropriate correction factors to obtain the actual values for the specified viscosity.

As a rule of thumb, centrifugal pumps should not be used for liquids whose viscosity is greater than 500 SSU. Figure 3.14 presents the effects of viscosity on centrifugal pump head and horsepower curves.

Since most pump performance characteristics are determined using water and are based on a specific gravity of 1.0, it is important to know the specific gravity of the liquid to be pumped so that the head required by a specific application can be determined. Figure 3.15 shows the relationship between specific gravity and head (energy) produced by the pump to overcome the head (energy) required by the system, and the horsepower required to deliver the specified flow rate and head (energy).

Effects of S.G. on performance

$$\text{Head (ft)} = \frac{(P_2 - P_1)\, 2.31}{\text{S.G.}}$$

$$\text{Horsepower} = \frac{\text{GPM} \times \text{head (ft)} \times \text{S.G.}}{3960 \times \text{efficiency}}$$

$$P_{DISCH} = \frac{\text{head (ft)} \times \text{S.G.} + P_1}{2.31}$$

Figure 3.15 Effects of S.G. on performance

Also, once the performance of a pump is known, the discharge pressure and horsepower will be influenced by changes in specific gravity of the liquid being pumped. Figure 3.16 presents a summary of centrifugal pump characteristics.

Pump characteristics
Kinetic (centrifugal)

- Variable flow
- Fixed head (differential pressure) at flow condition
- Limiting
- Throughput varies with S.G. change

Figure 3.16 Pump characteristics kinetic (centrifugal)

The effect of system resistance on performance

Now that centrifugal pump performance characteristics have been defined, we will consider the requirements of the system in which the pump is installed.

A system is a set of connected things or parts that work together. It is comprised of such things as vessels, pumps, valves, piping, exchangers, etc. which add to the total friction loss between the suction source and

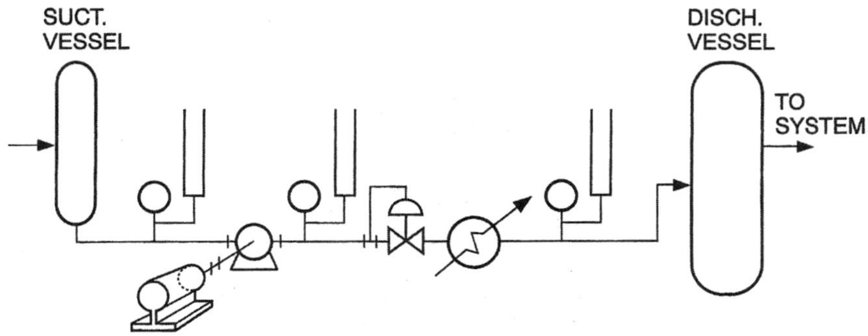

Figure 3.17 Simple systems

the final discharge point (refer to Figure 3.17 for a simple system).

As we know, the friction loss through the system is proportional to the square of the capacity (or velocity). This friction loss can be plotted as a curve overlaid on the pump performance curve. The intersection of the system curve and performance curve defines the operating point. Since a centrifugal pump is a variable flow, fixed head (for certain flow) device, any change in head required as a result of an increase in friction, elevation or ΔP in the vessels, will adjust the operating point back up the curve as appropriate to establish a new operating point. Figure 3.18 shows the effect of increased system resistance (head required) and the change of the operating point.

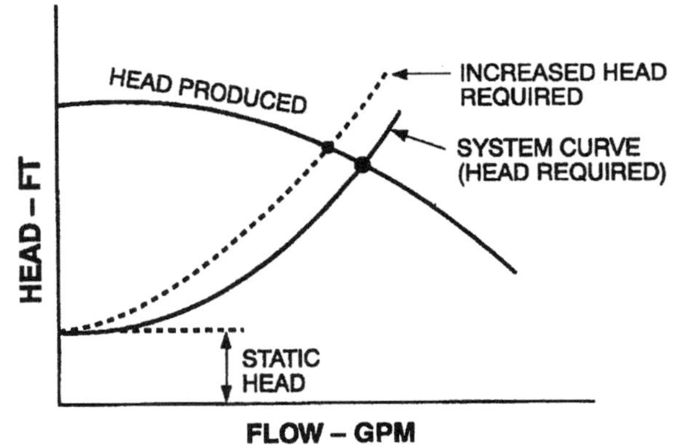

Figure 3.18 The effect of increased system resistance and the change of the operating point

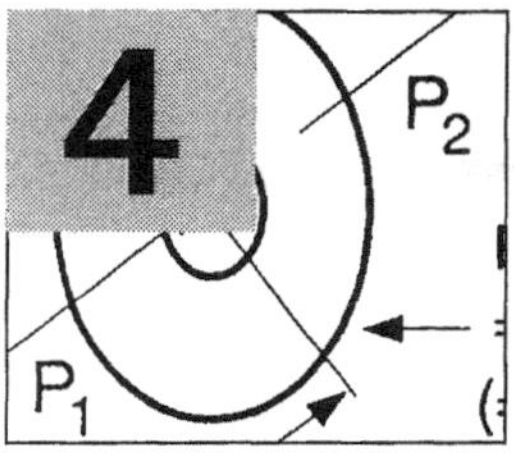

4 Operation of a pump in a process system

- Introduction
- A typical process system
- System pressure drop
- Static and head pressure
- A positive displacement pump in the process system
- A centrifugal pump in the process system
- Centrifugal pump stability

Introduction

Pumps, like compressors are an integral part of almost every process system. They need to operate continuously at a high level of reliability and availability in order to achieve established plant production quotas with minimum power usage and without loss of revenue. The effect of the process system on pump availability is significantly different for systems using kinetic pumps or positive displacement pumps. Kinetic pump capacity varies as the head (energy) required by the process system (ΔP at pump flanges and/or specific gravity) changes. Positive displacement pump capacity is not affected by process system head (energy) requirements. In this session we will examine the process system and the effects of that system on both classifications of pumps.

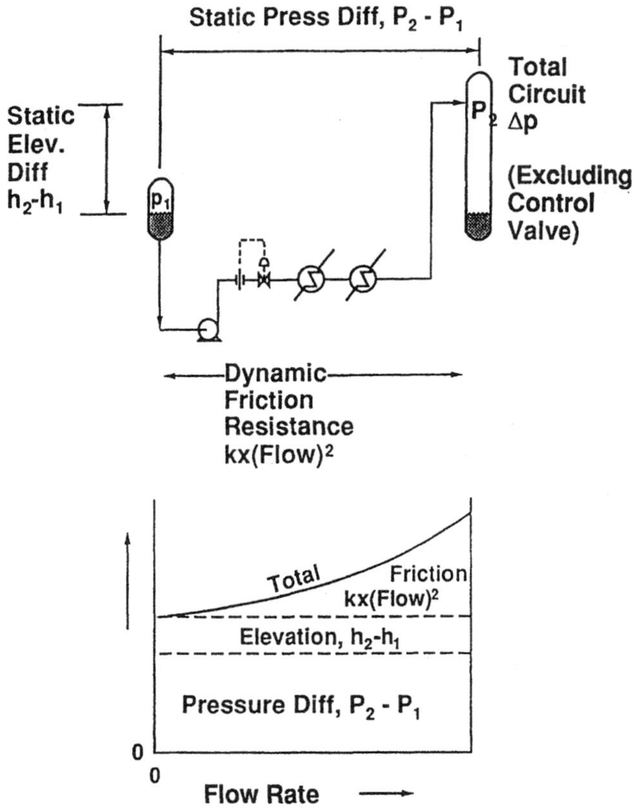

Figure 4.1 The system resistance curve

A typical process system

Pumps are required to move liquids from one pressure level to a higher level of pressure so that appropriate process steps can be undertaken. A typical process system is comprised of two parts; the suction system and the discharge system. It can include suction strainers, pressure vessels, pumps, furnaces, heat exchangers and a control valve (refer to Figure 4.1). All of these components add to the total friction loss in the system and define the operating point of flow and pressure for a pump handling a specific fluid at a given speed.

System pressure drop

The concept of an equivalent orifice will be introduced in this section (refer to Figure 4.2). Any pump impeller can be reduced to a series of equivalent orifices. The inlet of the impeller being one orifice, the exit

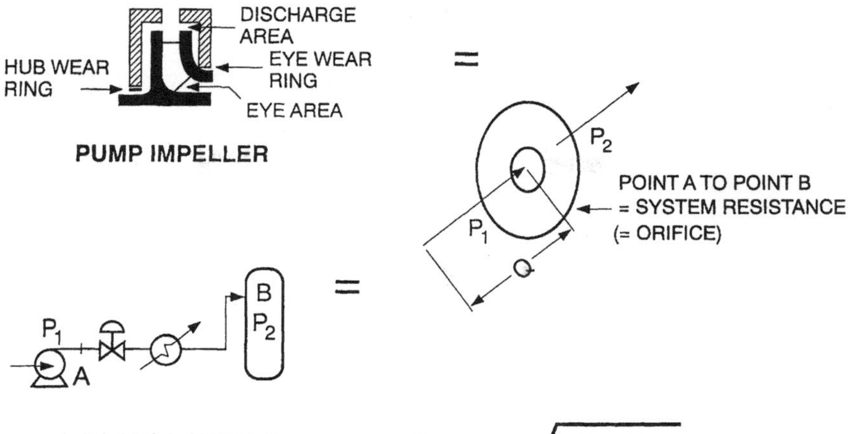

$$Q \propto d\sqrt{\frac{P_1 - P_2}{SG}}$$

Figure 4.2 Concept of an equivalent orifice

being another, and the impeller case wear ring clearances being additional equivalent orifices. It should also be noted that both the suction and discharge process system for a given point in time and flow rate can be thought of as equivalent orifices placed at the suction and discharge flanges of the pump.

To understand how a pump performs in a process system, we will examine various system resistance curves, and define the operating point of the pump within a typical process system (a set of connected things that work together).

By examining the system in Figure 4.1, it can be seen that any system relative to the pump is comprised of two parts; the suction system and the discharge system. Therefore, the objective of the pump is to move liquid from the suction system process pressure level to the final discharge pressure level. Liquid flows from the suction vessel to the pump as a result of a reduction in pressure resulting from friction loss in this part of the system (above vapor pressure to present flashing). The discharge pressure at the pump flange is an accumulation of the system resistance (piping, exchanger, furnace, control valve, etc.) from the terminal point back to the pump flange. An important point to remember is that pump differential head or energy required is the net effect of the discharge system resistance and the suction system resistance.

Static and head pressure

In addition to the friction loss component, the system head or resistance has two other components; static pressure and elevation head. Friction loss varies as the square of the flow rate and depends on the piping layout, pipe diameter and equivalent length (including valves and fittings). Static pressure is the difference in pressure of the discharge pressure vessel and the suction pressure vessel. Elevation head is the difference in liquid level between the suction and discharge vessels. These three components are totalled and plotted against flow rate to create a system head required curve (refer to Figure 4.1).

If we review the previous section and examine the operating characteristics of both positive displacement and dynamic pumps in a simple system, two important facts concerning the performance of these pumps becomes evident.

A positive displacement pump in the process system

A positive displacement pump will produce nearly constant flow while continually adding head (energy) to the liquid provided sufficient power is available and the pump and drive system is designed to meet this objective. Therefore, a positive displacement pump will be relatively insensitive to both system changes and specific gravity changes (refer to Figure 4.3).

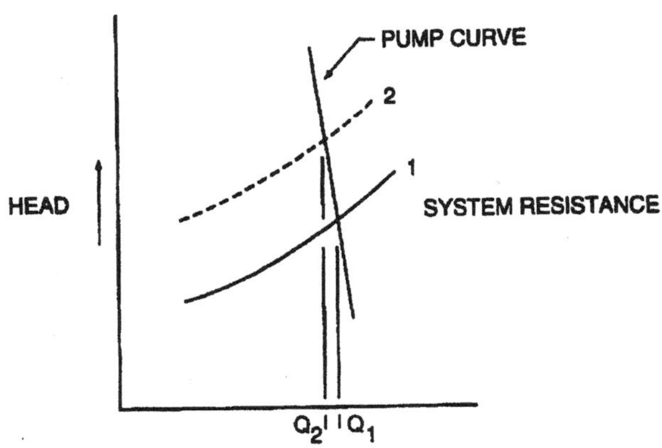

Figure 4.3 A positive displacement pump in a process system

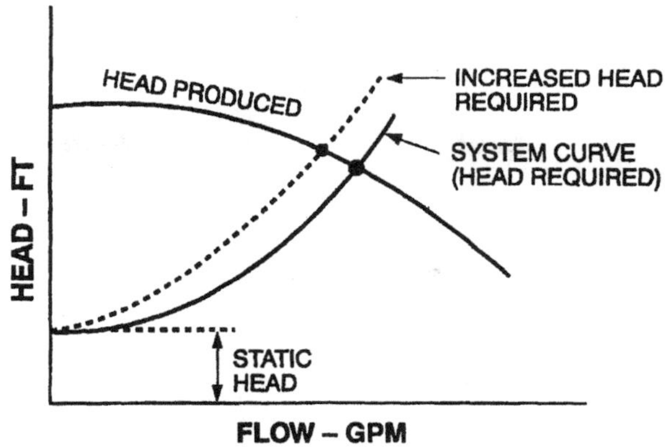

Figure 4.4 A centrifugal pump in a process system

A centrifugal pump in the process system

On the other hand, since the characteristic of the centrifugal (kinetic) pump is to add head (energy) to the liquid by working on the liquid with rotating blades, velocities play an important part in the make up of this pump's characteristics (refer to Figure 4.4). Anything that will result in a velocity change at the tip of the blade (impeller) will result in a change in produced head (energy) and a corresponding flow change. This type of pump will be extremely sensitive to system resistance changes or S.G. since an increase on system resistance requirement will force the pump to operate at a reduced flow rate.

Centrifugal pump stability

An additional point to understand is that the head (energy) required by the process system is a function of the system pressure drop and the specific gravity of the liquid. Any change in either of these variables will change the head (energy) required by the process and affect the operating point of a kinetic pump. The operating point of positive displacement pump will not be affected since it has infinite head (energy) capability.

As stated earlier in this section, head (energy) required is a result of the difference in static pressure and elevation head, and friction loss which is plotted against flow rate. Different process systems have different head (energy) required curve shapes (refer to Figure 4.5).

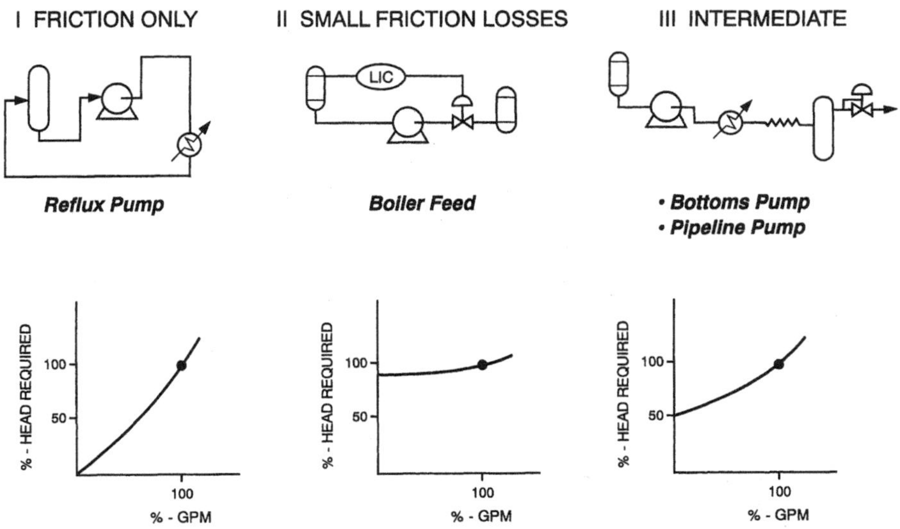

Figure 4.5 System resistance characteristics

A reflux loop for example, is comprised of friction loss only and will have a relatively steep system resistance curve, whereas a boiler feed pump with a relative small system resistance, will have a system resistance curve which is less steep. One can see that the combination of a relatively flat system curve and pump characteristic curve which has the shutoff head lower than the maximum head, can lead to unstable operation. The most common system characteristic is the intermediate case – examples being bottoms pump and pipeline pump.

It is a common misconception that a pump with a flat head curve is inherently unstable. This is not necessarily true since the operating point of any pump is the equilibrium point of the head (energy) required by the process and the head (energy) produced by the pump. Even though the pump head curve is flat, a rising head required curve (system resistance) will result in stable pump operation. Refer again to Figure 4.5. It can be seen from this figure that a flat pump head curve is to be avoided in a process system that has a flat head required (system resistance) curve.

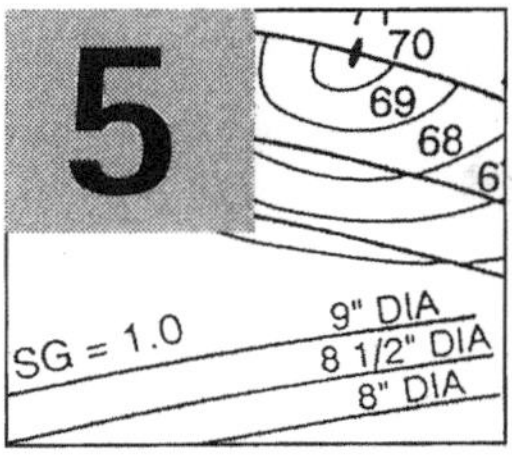

5 Pump performance data

- The pump curve
- The limits of the centrifugal pump curve
- Increasing head produced by a centrifugal pump

The pump curve

The pump characteristic curve defines the signature of a pump over its entire life. It identifies the range of flows and energy produced for a fixed speed, size, design and suction conditions of the pump. In order for a pump to move a liquid from a level of low pressure to a level of higher pressure, the head (energy) produced by the pump must equal or exceed the net head (energy) required by the system. Also, the net head (energy) available on the suction side of the pump system must be greater than the liquid vapor pressure to assure that liquid enters the pump without potential deterioration of performance or mechanical damage.

The positive displacement curve

Figure 5.1 illustrates the characteristics which defines the performance curve for a rotary type positive displacement pump.

- Capacity, GPM = Displacement – Slip
- Pressure, PSI
- Brake HP = Friction HP + Hydraulic HP
- EFF $= \dfrac{\text{GPM x PSID}}{\text{1714 x BHP}}$

$$= \frac{\text{GPM x HEAD x S.G.}}{\text{3960 x BHP}}$$

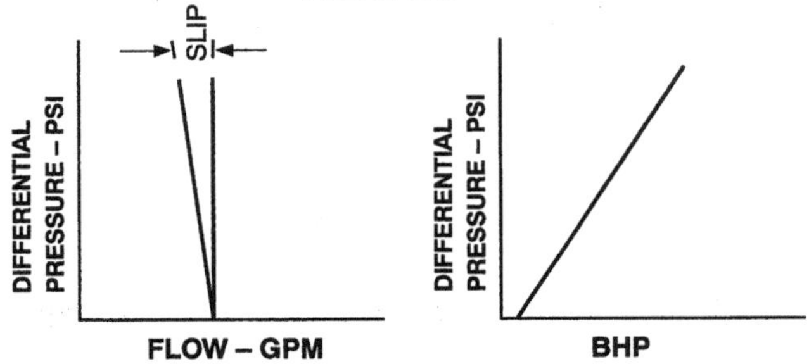

Figure 5.1 The positive displacement pump performance

Head

As shown in Figure 5.1, differential pressure is usually plotted vs flow instead of head to express energy produced by positive displacement pumps. Since the head (energy) produced is infinite for all types of P.D. pumps, produced differential pressure is not affected by specific gravity as in the case of centrifugal pumps, therefore differential pressure can be used to express the head (energy) produced by P.D. pumps.

Flow

As shown in Figure 5.1, flow reduces slightly with increased energy produced. This is a result of increased pump slip.

Efficiency

Positive displacement pump efficiencies are expressed as volumetric efficiency which is defined as:

$$\text{V.E.} = \frac{\text{actual capacity}}{\text{displacement}}$$

The displacement is defined as the capacity the pump would produce if there were no (slip) leakage losses. P.D. pump efficiencies vary from:

- 70% – 80% for rotary pumps
- 75% – 85% for plunger and reciprocating pumps

Horsepower

As shown in Figure 5.1 horsepower is directly proportional to pump differential pressure for a given liquid.

NPSH required

To overcome the continuous action of pulsating flow, which occurs in reciprocating type P.D. pumps, the acceleration energy must also be accounted for in the system to assure that the liquid to be pumped does not vaporize prior to entering the pump. It should be noted that "acceleration head loss" is the term used to express this energy. Refer to Figure 5.2 for the equations used to calculate the NPSH available for a system with and without pulsating flow characteristics.

NPSH available calculation for a PD pump

$$NPSH_{AVAIL} = (P_{S1} + P_A)\,\frac{2.31}{S.G.} + H_{Z1} - H_F - (P_V)\,\frac{2.31}{S.G.} \text{ (non pulsating flow)}$$

$$NPSH_{AVAIL} = (P_{S1} + P_A)\,\frac{2.31}{S.G.} + H - H_F - H_{LA} - (P_V)\,\frac{2.31}{S.G.} \text{ (pulsating flow)}$$

P_{S1} = suction press. PSIG	H_{Z1} = suction side elevation change
P_A = atmos. press. PSIG	H_F = suction side system friction (expressed in FT.)
S.G. = specific gravity	H_{LA} = acceleration head loss (magnitude depends on pump type)
P_V = vapor pressure	PSIA

Figure 5.2 NPSH available calculation for a PD pump

The centrifugal curve

When a centrifugal pump is designed, its performance characteristics are defined for a range of flows and head (energy) produced for fixed impeller geometry and a variety of impeller diameters. Figures 5.3 and 5.4 show the relationships between the head (energy) produced and the flow through a single stage centrifugal pump. By the affinity laws, pumps of the same type, fitted with impellers of similar design, will have characteristics curves of the same shape.

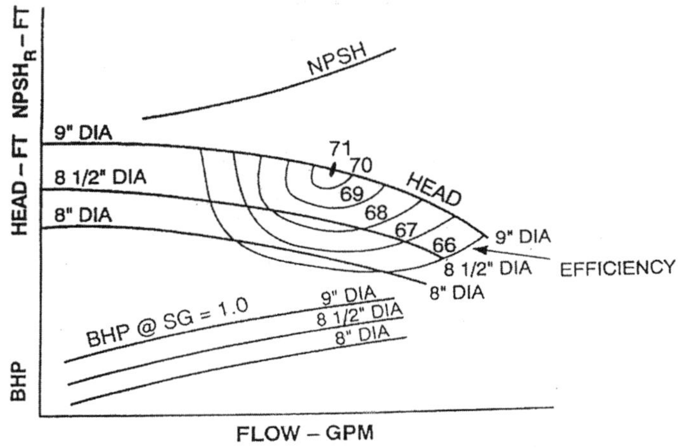

Figure 5.3 A typical centrifugal pump performance curve

Centrifugal pump performance definitions

$$H = \frac{V^2}{2g}$$

H = energy expressed in feet of liquid

V = velocity in feet per second

g = acceleration due to gravity in FT/SEC^2

$$BHP = \frac{GPM \times H \times S.G.}{3960 \times EFF}$$

EFF ratio of power output to power input

NPSHR energy which must be exceeded to avoid vaporization of liquid in pump suction passage

Figure 5.4 Centrifugal pump performance definitions

Head produced

The head produced by a centrifugal pump varies inversely with the flow rate. The curve head rise is a function of the impeller inlet and discharge blade angles. Typical centrifugal pump head rise values from design point to shutoff are 5% – 15%.

When the head required by the process exceeds the head produced by a single stage centrifugal pump, multistaging is used to produce the energy required by the system. Multistaging is nothing more than two or more impellers acting in series within a single casing to produce the total head (energy) required. It is common practice for each impeller to produce an equal amount of energy (refer to Figure 5.5).

Example – multistaging

- total net system energy (ft. head) - 1200 ft.
- number of impellers selected – 4
- energy (ft. head) produced for impeller – $\frac{1200}{4}$ = 300 ft/impeller

Figure 5.5 Example – multistaging

Flow

The flow rate of a centrifugal pump varies inversely with the head (energy) required by the process. For a given impeller design operating at a constant speed, increased process head requirements will reduce centrifugal pump flow rates. Since the typical head rise values for centrifugal pumps are 5% – 15%, a relatively small change in process head requirements can result in significant flow reductions and possible impeller recirculation on operation near zero flow (shutoff).

Efficiency

The pump efficiency is maximum at the pump design point using the maximum diameter impeller. Refer to Figure 5.3. The pump design point, often referred to as the B.E.P. – best efficiency point, is the flow where minimum losses occur in the pump stationary passages and the impeller. At off design flows, separation losses (low flows) and turbulence losses (high flows) increase internal produced head losses and reduce pump efficiency.

Horsepower

Horsepower required by a centrifugal pump varies directly with the specific gravity of the pumped liquid. Horsepower is the only parameter on a typical centrifugal pump curve that is effected by the specific gravity of the pumped liquid. Most pump curves present the horsepower curve based on water S.G. = 1.0. For pumped liquids of any other S.G. value, the horsepower on the pump curve must be multiplied by the actual specific gravity.

$NPSH_R$

The net positive suction head required by a centrifugal pump varies approximately with the flow rate squared since it is a measure of the pressure drop from the pump suction flange to the eye of the first impeller.

The $NPSH_R$ is also influenced by the pump rotational speed and varies somewhat less than the rotational speed squared.

The limits of the centrifugal pump curve

The centrifugal pump curve has high flow and low flow limits which can cause significant mechanical damage to the pump if not avoided. At the low flow end of the curve, flow recirculation can damage a pump while at the high flow end excessive $NPSH_{REQUIRED}$, horsepower and choke flow can result in mechanical damage to impellers, casing, shaft, bearings and seals. Each of these factors is discussed below.

Low flow operation

As we examine these factors we can see that oversizing a centrifugal pump will result in low flows through the impeller. A portion of the flow will reverse itself and set up turbulence as it reenters the impeller. The abrupt change in direction and very high acceleration can result in cavitation on the back side of the impeller vane (refer to Figure 5.6).

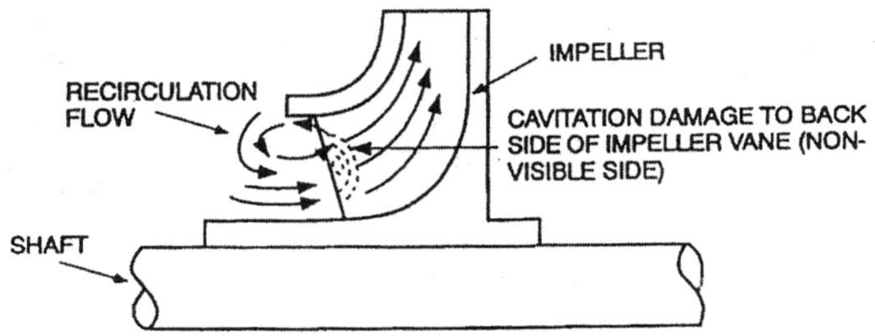

Figure 5.6 Recirculation flow pattern in impeller at low flows

As a result of oversizing an impeller, significant effects in performance and mechanical reliability can be experienced. These are outlined in Figure 5.7.

Effects of pump oversizing

Operation at low flows can result in:

- Internal recirculation damage to impeller
- Operation at less than best efficiency point
- High radial loads
- Bearing failures
- Seal failures
- High internal temperature rise and requirement for minimum flow bypass

Figure 5.7 Effects of pump oversizing

Pumps are designed to operate at minimum radial thrust loads at best efficiency point. Low flow operation results in high radial loads which can cause premature bearing failures unless bearings are selected to accept these higher loads in anticipation of operation at low flows. Pressure surges and flashing of the liquid can also occur at low flows. This can cause loading and unloading of the mechanical seal faces which can result in a seal failure. Depending on the fluid being pumped, low flow operation can result in a high temperature rise through the pump because the amount of energy absorbed by the liquid is low compared to that absorbed by friction losses. Refer to Figure 5.8 for the method to calculate the temperature rise through a pump.

Temperature rise through a pump

$$\text{RISE, DEG F} = \frac{H}{778 \times C_P} \left[\frac{1}{\text{EFF'Y}} - 1\right]$$

H = head in feet at pumping rate

eff'y = efficiency at pumping rate

CP = specific heat, BTU/LB–°F

778 = FT. LB/BTU

Figure 5.8 Temperature rise through a pump

The above relationship can also be used to determine the approximate flow rate of any centrifugal pump. By measuring the pipe temperature rise across any pump, the efficiency of the pump can be determined. Referring to the particular pump shop test curve for the calculated efficiency will allow the approximate pump flow rate to be determined. Note: This approach assumes the pump is in new condition. A worn condition will reduce the flow to a greater extent.

High flow operation

Selecting a pump to operate far to the right of best efficiency point can also result in potential pump problems as highlighted in Figure 5.9.

Effects of pump operation at high flows

Operation at high flows can result in:

- High to overloading horsepower with reduced system resistance
- Operation in the "break" of head capacity curve (significant changes in head with no change in flows)
- Higher NPSH required than available
- Recirculation cavitation at impeller tips

Figure 5.9 Effects of pump operation at high flows

Pump curve shapes

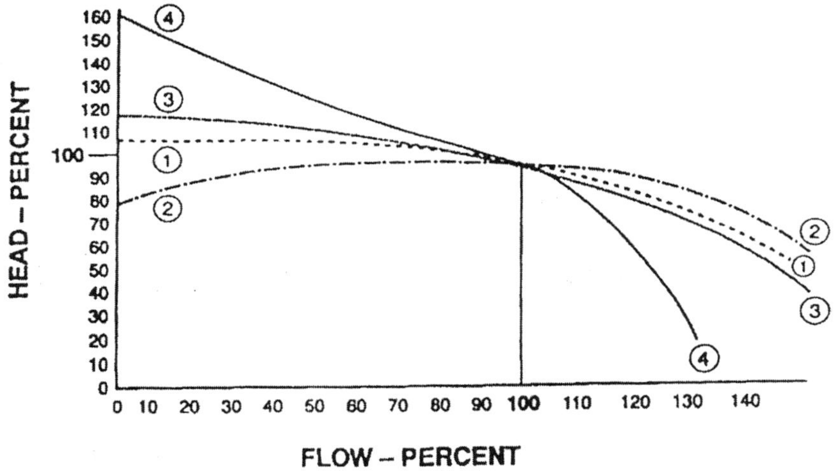

Figure 5.10 Centrifugal pump performance curve shapes

The various types of characteristic curves normally associated with centrifugal pumps include flat, drooping, rising, stable and unstable depending upon their shape. Figure 5.10 illustrates the different curve shapes and Figure 5.11 defines each type. The pump curve shape can play a significant role in determining if stable operation in a given process system is possible. Flat or drooping head characteristic curves (Figure 5.10 – curves 1 & 2) can result in unstable operation (varying flow rates). Pumps should be selected with a rising head curve or controlled such that they always operate in the rising region of their curve.

Definition of characteristic curve shape

Flat curves (1 and 2), show little variation in head for all flows between design point and shut-off

Drooping curve (2), head at shut-off is less than head developed at some flows between design point to shut-off

Rising curve (1,3,4), head rises continuously as flow decreases from design point to shut-off

Steep curve (4), large increase between head developed at design flow and that developed at shut-off

Stable curve (1,3,4), one flow rate for any one head

Unstable curve (2), same head can be developed at more than one flow rate

Figure 5.11 Definition of characteristic curve shape

Increasing head produced by a centrifugal pump

The affinity laws can be used to increase the head available from a centrifugal pump. Head produced by a centrifugal pump is a function of impeller tip speed. Since tip speed is a function of impeller diameter and rotational speed, two options are available. The characteristic curve can be affected by either a speed change or a change in impeller diameter with speed held constant. Figures 5.12 A&B show this relationship.

The affinity laws

Increasing head produced by a pump

$$H_1 = \left(\frac{D_1}{D}\right)^2 H \text{ or } H_1 = \left(\frac{N_1}{N}\right)^2 H$$

where: D_1 = diameter of changed impeller

D = diameter of existing impeller

H_1 = head in feet changed impeller

H = head in feet existing impeller

N_1 = Speed (RPM) changed condition

N = speed (RPM) existing condition

Figure 5.12A The affinity laws

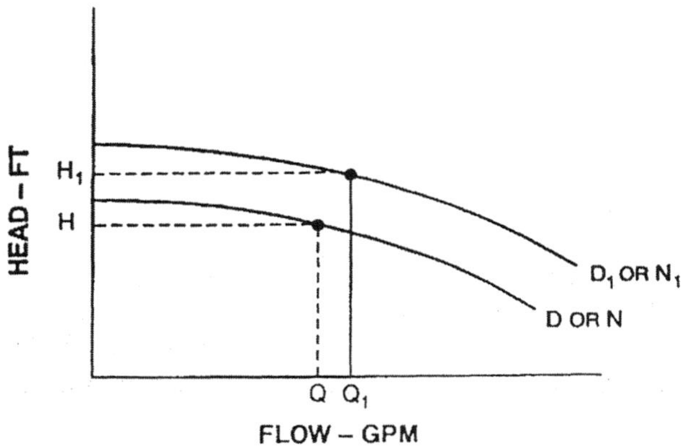

Figure 5.12B The affinity laws

The affinity laws

In actual practice, the affinity laws provide an approximation between flow, head and horsepower as pump impeller diameter or speed is varied. The actual values vary somewhat less than predicted by the affinity laws. That is, the actual exponents in the affinity equations are slightly less than their stated values and are different for each pump. This fact is a result of friction in hydraulic passages and impellers, leakage losses and variation of impeller discharge vane angles when diameters are changed. Pump manufacturers should be contacted to confirm actual impeller diameters and speed changes to meet new duty requirements.

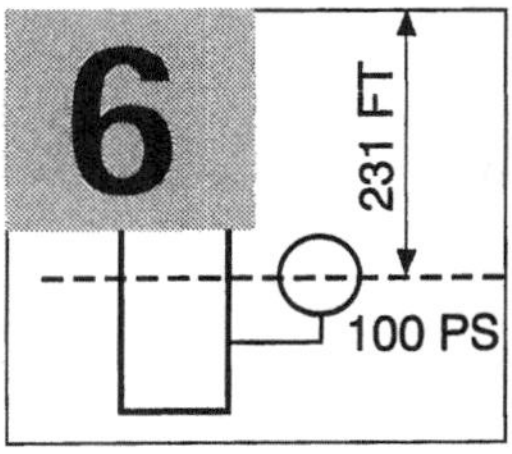

The concept of pump head

- Definition of pump head
- Process system head requirements
- Head available (produced)
- The operating point

Definition of pump head

In simple terms head is the energy in foot pounds to move one pound of fluid from one energy level to another. Industry however, has always defined head in feet. Head should be expressed in foot pounds force per pound mass or British Thermal Units per pound. A British Thermal Unit is equal to exactly 778 foot pounds force per pound mass of that fluid. Refer to Figure 6.1.

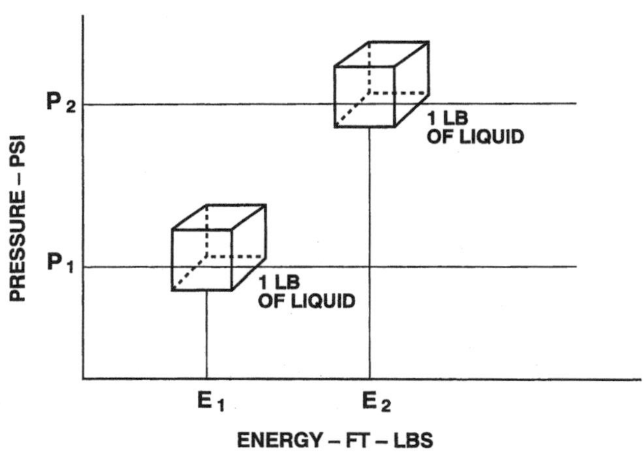

Figure 6.1 Head (energy) for liquid

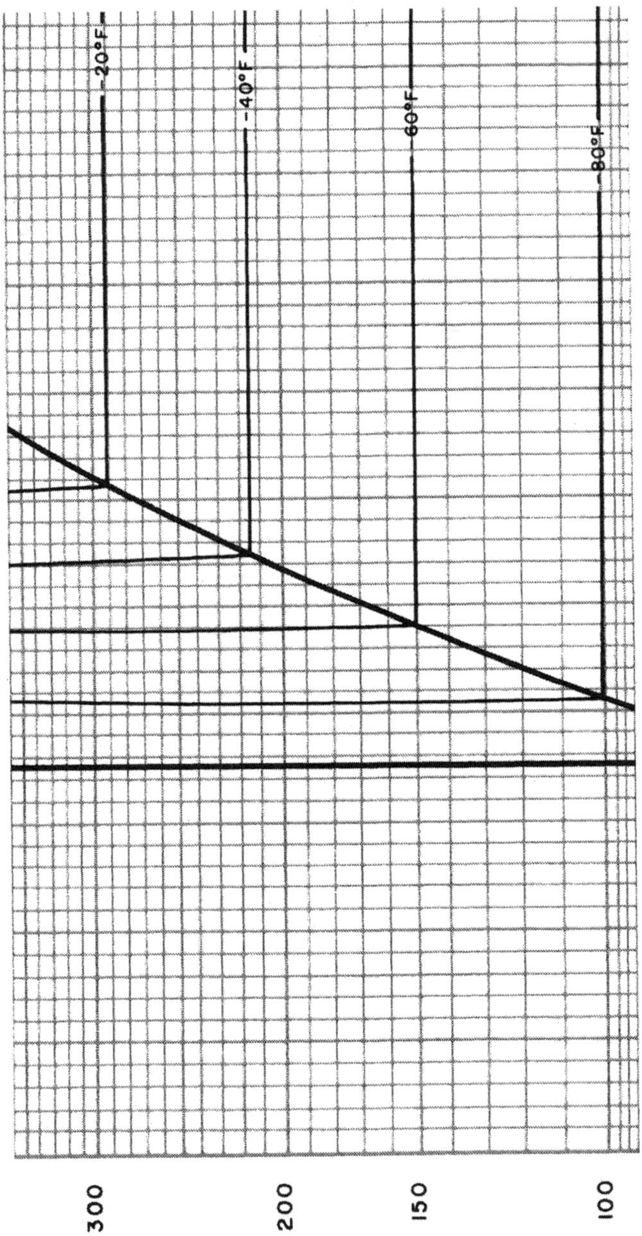

Figure 6.2 Ethylene Mollier diagram

When we deal with fluid head, a fluid can be either a liquid or a gas depending upon the state of the fluid at that specific time. Ethylene for example, can be either a liquid or gas depending on its pressure and temperature. If it is a liquid, an ethylene pulp will be used and the energy required to increase the pressure of the liquid from P_1 to P_2 will be defined as head, in foot pounds force per pound mass. Conversely, if the fluid conditions were such that it would be a vapor, then are ethylene compressor would be used to achieve the same result (refer to Figure 6.2).

Process system head requirements

It can also be stated that head is the energy required to accomplish a specific task within a defined process system. The amount of head (energy) to move a liquid from one pressure level to another varies directly with velocity (system friction loss increases) and inversely with specific gravity (refer to Figure 6.3).

Effects of specific gravity on liquid head

$$H = \frac{2.31 \times \Delta P}{S.G.}$$

Water	**Naphtha**
$\Delta P = 100$ psi	$\Delta P = 100$ psi
S.G. = 1.0	S.G. = .78
$P_1 = 14.7$ psia	$P_1 = 14.7$ psia
$P_2 = 114.7$ psia	$P_2 = 114.7$ psia
$H = 231\ ft.lb_f/lb_m$	$H = 296\ ft.lb_f/lb_m$

ΔP = Differential pressure and includes system friction loss (velocity effects)

Figure 6.3 Effects of specific gravity on liquid head

Another way to illustrate the relationship of head required as a function of the process and liquid composition, is to compare liquid columns of water, caustic and naphtha to effect the same level of pressure (refer to Figure 6.4). As can be seen from Figure 6.4, head required is inversely proportional to fluid mass or specific gravity.

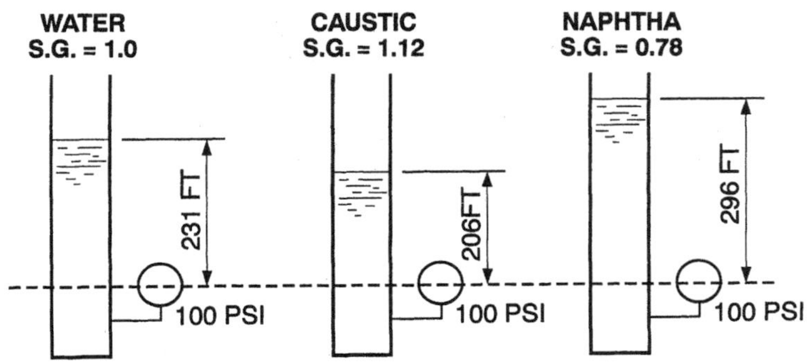

Figure 6.4 Effect of liquid composition on head (energy)

Head available (produced)

In order to overcome the head (energy) required by the process system, a pump must produce a certain amount of head (energy) which is equal to or greater than that required by the system. Changes to the composition of the liquid will not affect the head (energy) produced by a pump. However, discharge pressure will definitely be affected by composition changes. Therefore, head (energy) versus flow rate should be used to describe pump performance. The factors which determine the head (energy) produced by a pump are described in Figure 6.5.

Head (energy) produced

- head (energy) produced is determined by:
  - pump impeller design
- head produced by impeller:
  - increases with tip speed
    - impeller diameter
    - pump speed (rpm)
  - increases with decreasing flow

Figure 6.5 Head (energy) produced

To understand the concept of how head (energy) produced changes with tip speed and flowrate, it is necessary to describe what is actually happening to the liquid as the impeller rotates within the pump. The rotating motion of the impeller results in two components to the motion imparted to the liquid. One motion is radial in direction from the centre of the impeller and the other is tangential to the outside

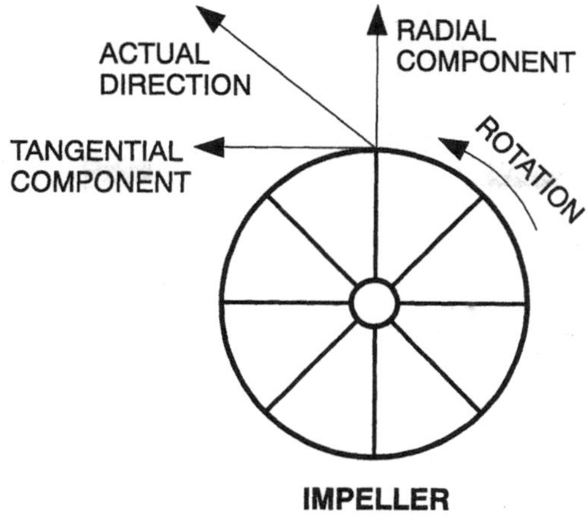

Figure 6.6 Components of motion produced by a pump impeller

diameter of the impeller. The actual direction is the resultant of the two flow directions (refer to Figure 6.6).

In addition, it is important to understand the concept of an equivalent orifice. Given any impeller configuration, specific areas can be reduced to equivalent orifices, namely; the inlet or eye area, discharge area between any two vanes, the eye seal and hub seal (refer to Figure 6.7). This concept makes it much easier to understand that for a given area, flow will change as a square root function of the differential pressure.

If we examine the effect of vane angle, we will see what happens when the flow changes from a point of rated flow to a lower flow. Let us

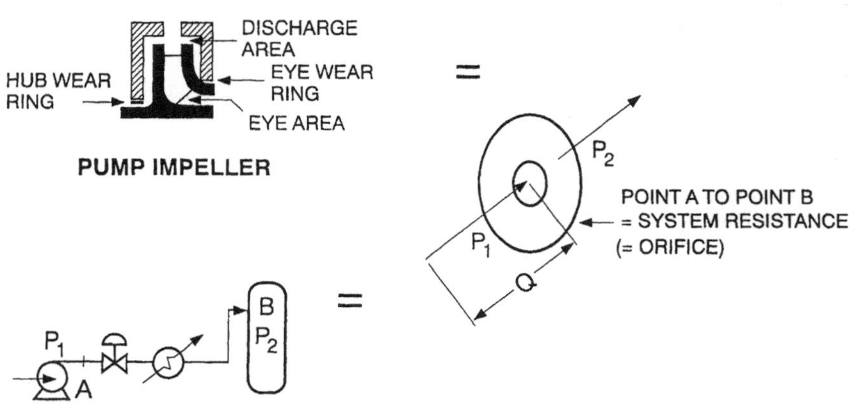

$$Q \propto d \sqrt{\frac{P_1 - P_2}{SG}}$$

Figure 6.7 Concept of an equivalent orifice

assume that the impeller has radial vanes. At the rated point, the velocity relative to the vane is completely radial, assuming zero slip, and tip speed is defined by the impeller diameter and speed (rpm) (refer to Figure 6.8 for important relationships).

Impeller flow relationships

- Vector describes magnitude and direction
- Tip speed 'U' = $\frac{DN}{229}$
- Flow related to velocity 'Q' = (A) (V) 500

Where U = tip velocity (ft./sec.) Q = flow rate (gal/min)

D = diameter (in) A = area (ft^2)

N = speed (rpm) V = velocity (ft./sec.)

Figure 6.8 Impeller flow relationships

For radial vaned impellers, the absolute velocity is the sum of the two vectors (relative and tip velocity). Project the tangential velocity (X-axis projection) from the absolute velocity and note its value.

At a lower flow, the tip speed will remain constant (assuming constant speed) and the relative velocity follows the radial vane path, the magnitude of the tangential velocity remains constant regardless of the value of the relative velocity. Since energy produced by the vane is the product of tip speed (unchanged) and the tangential velocity (unchanged) the head (energy) produced in the radial vaned impeller will remain essentially the same. Therefore, the curve shape will be significantly flatter than for a non radial vaned impeller (Figure 6.9 illustrates the two velocities generated by a radial vaned impeller). In reality however, the effects of friction will produce a curve shape that will increase from high flows to low flows but the effects will produce less of an energy increase than for a non radial vane design. A typical head rise from design point to shutoff is less than 5% for a radial vaned impeller.

This is an important fact to remember since the operating point of any dynamic pump will be the intersection of the system resistance (head required) and the pump performance curve (head produced). It should be noted that regardless of the type of liquid used in pumps, velocity relative to the vane will not change for different fluids at a constant flow

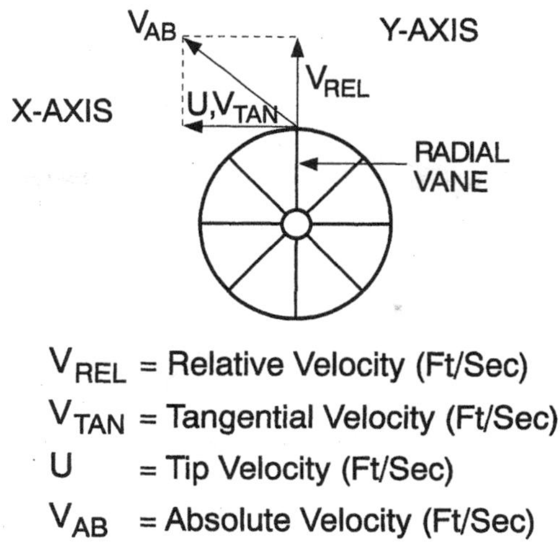

Figure 6.9 Velocities generated by radial vanes

rate into the impeller since liquids are incompressible. Therefore, for a given pump impeller design operating at a constant flow and speed, the head produced will be constant regardless of the liquid specific gravity assuming viscosity effects are negligible.

Let us now examine the flow pattern between any two vanes of a non radial vaned impeller design which is the most common pump impeller design. The blade tip velocity 'U' is a function of the diameter of the vane and the vane rotational speed. The velocity relative to the vane is a function of the area between the vanes, the flow rate in that area and the angle of the vane at the discharge of the impeller. Summing these two velocities, the resultant or absolute velocity defines the magnitude and direction of the liquid as it exits the vane. For this discussion, we assume that the velocity relative to the vane exactly follows the vane angles, that is, slip is zero. This assumption can be used since it will not impact the final conclusion of our discussion. If we now resolve the absolute velocity into X and Y components we see that the X projection of the component is the tangential velocity of the liquid relative to the impeller. The head (energy) produced by any dynamic machine is proportional to the product of the tip velocity and the tangential velocity (refer to Figure 6.10).

The operating point

The operating point is the equilibrium point between the head produced by any type pump and the head required by the process in which it operates.

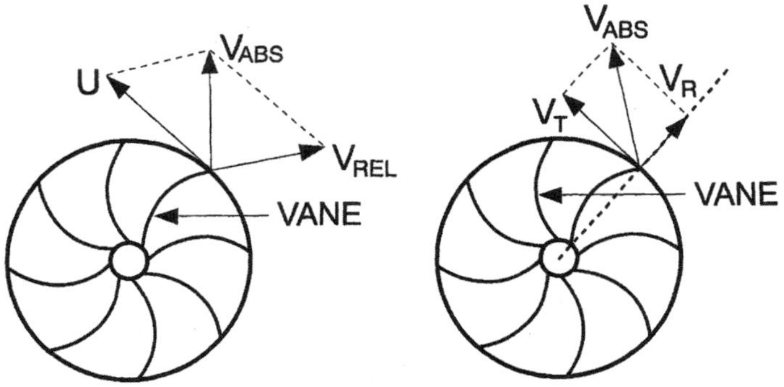

Figure 6.10 Velocities generated by non radial vanes

Centrifugal pumps

If we now change the system resistance such that the flow through the impeller reduces, we can examine the discharge velocity triangle to see what changes have occurred. Assuming the speed is constant, we can see that the tip velocity does not change. However, the velocity relative to the vane will be reduced as a result of less flow passing through a fixed area between the vanes. If we sum the velocity vectors to obtain the absolute velocity, we can see that the angle of the liquid leaving the vane is significantly reduced and the X projection of the tangential velocity will be greater than the value at the original rated point. Since the energy produced by the vane is proportional to the tip speed (unchanged) and the tangential velocity (increased), we can see that a reduction in flow through the vanes has resulted in increased head or energy imparted to the fluid. This makes sense since the slower the liquid proceeds through the vane, the more time it has to pick up energy imparted by the vanes, and as a result, increases the energy level. Therefore, we can see that for all dynamic vanes and impellers which increase energy of a fluid by the action of the vanes, increased fluid energy can only occur at a lower flow assuming the speed and the initial angle of fluid relative to the vane remains unchanged.

Figure 6.11 illustrates the effect of a shifting operating point for a non radial vaned centrifugal pump impeller, as result of increased system resistance (head required). It also reinforces the concept that increased (head) energy produced occurs only as flow is reduced (with speed constant).

The increased produced head (energy) occurs as a result of lower velocity through a fixed area (between impeller vanes) and longer contact time of the liquid relative to the vane. It must be remembered that the operating point is the equilibrium point where the head (energy) required is equal to the head (energy) produced from either a dynamic or positive displacement pump. Figure 6.12 shows the effect

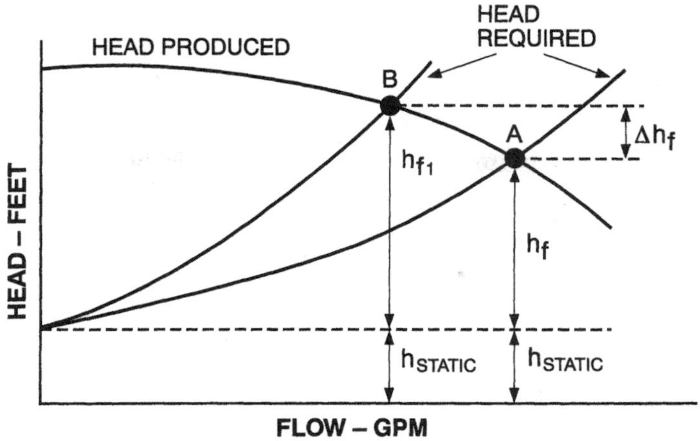

Figure 6.11 Shifting operating point for non radial vaned impeller

of a shifting operating point for a radial vaned centrifugal pump impeller. Note that there is no significant increase in head (energy) produced with increased head (energy) required by the process system. This occurs because of the flatness of the performance curve generated by the radial vane impeller.

Positive displacement pumps

The characteristic of any positive displacement pump when plotted against head produced and flow coordinates is essentially a vertical line (neglecting slip). See Figure 6.13. Stated in terms of characteristics, any

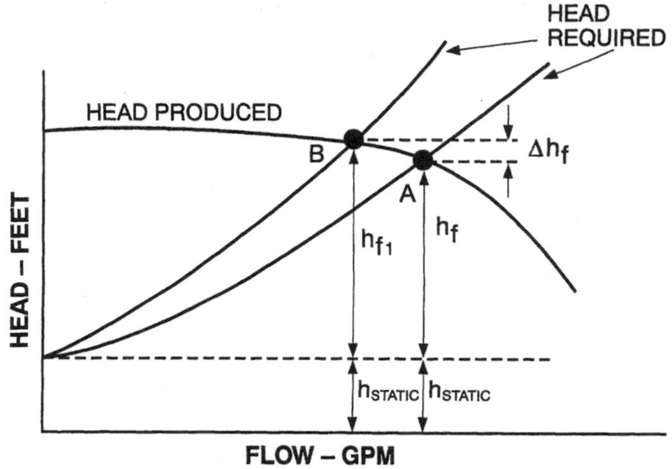

Figure 6.12 Shifting operating point for radial vaned impeller

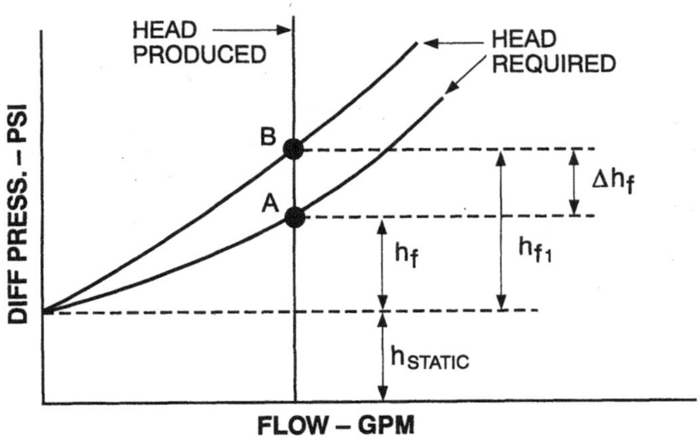

Figure 6.13 Shifting operating point for a positive displacement pump

positive displacement pump is a constant flow device that possesses infinite head produced capability. As a result, any change in head required by the process will not significantly effect flow rate.

7 Pump selection

- Introduction
- Start with a data sheet to completely define requirements
- Completely define the operating conditions
- Defining the pump rated point for efficient operation
- Carefully define critical component requirements
- Guidelines to use when selecting pump style
- Pump selection examples

Introduction

This chapter will cover the major principles required for cost effective pump selection.

Start with a data sheet to completely define requirements

One of the single most important factors in selecting a pump to meet the requirements of a process system is to completely and accurately state all the requirements on a data sheet. A centrifugal pump data sheet, courtesy of the American Petroleum Industry (API 610) is supplied at the end of the Chapter 7.

Regardless of the source, all pump data sheets should contain the categories of information shown in Figure 7.1

Minimum data sheet requirements

(P) (U) ■ Pump application and operating mode (single or parallel)
(P) (U) ■ Detailed operating conditions
(P) (U) ■ Accurate site and utility requirements
(M) ■ Pump performance
(P) (M) ■ Pump construction and experience
(P) ■ Spare parts required
(P) (M) ■ Driver details
(P) ■ QA inspection and test requirements

Note: to be completed by:
P = Purchaser
M = Manufacturer (vendor)
U = User

Figure 7.1 Minimum data sheet requirements

Completely define the operating conditions

Correctly stated operating conditions are essential for proper definition and subsequent selection of a specific type and configuration of pump to meet the specified conditions.

Once it is decided to install a pumping system, a sketch should be drawn to define all of the components which are required to be included into that system. Some of the factors which need to be considered in completing the sketch and system design includes the following:

- Flow rate – All flow rates including minimum, normal and rated should be listed in the data sheet. Normal flow is usually the flow required to achieve a specific process operation. The rated flow is normally a set percentage increase over the normal flow and it usually includes consideration for pump wear and the type of operation within the process system. It can amount to as much as ten (10) percent depending upon specific company practice. The minimum flow is important to identify in order to establish if a minimum flow bypass line is required for process or mechanical design considerations.

- Head required – The required head that the pump must develop is based on the static pressure difference between the discharge terminal point and the suction source, the elevation difference and the friction losses through system components including suction and discharge side piping, pressure drop through heat exchangers, furnaces, control valves and other equipment. It is represented by the equation in Figure 7.2.

Required head equation

$$H = \frac{2.311 \times \Delta P \text{ (at pump flanges)}}{S.G.}$$

ΔP = Total pressure difference between the discharge system and suction system, measured at the pump flanges

S.G. = Specific gravity of the liquid at pump temperature

Figure 7.2 Required head equation

- Liquid properties – **Viscosity, vapor pressure and specific gravity** each play an important role in achieving the required level of pump reliability within the operating system. **Viscosity** can impact pump performance to the extent that it may not be justified to even use a centrifugal pump when the viscosity values are greater than 3000–5000 SSU. The hydraulic institute has published curves which can be used to calculate the performance effects resulting from pumping viscous liquid.

  Vapor pressure and **specific gravity** influence the type of pump to use and it's mechanical design configuration. **Vapor pressure** is an important property when determining whether there is adequate net energy available at he pump suction to avoid vaporization of the liquid which can lead to performance deterioration and possible shortened life expectancy of the pump.

  Specific gravity is the liquid property used to calculate the amount of head a pump has to produce to overcome the resistance of the suction and discharge systems. It is also used as a guideline to determine whether a pump casing design should be of the vertical (radial) split or horizontal (axial) configuration (refer to Figure 7.3 for some guidelines).

Casing configuration guidelines

Use radial split casing for:

- S.G. ≤ 0.7 @ pumping temperature
- Pumping temperature ≥ 400°F
- Flammable or toxic liquids at rated discharge pressures above 1000 psig

Figure 7.3 Casing configuration guidelines

- NPSH available – Net positive suction head available is a characteristic of the process suction system. It is the energy above the vapor pressure of the liquid, measured at the suction flange of the pump, which is required to maintain the fluid in a liquid state. In a centrifugal pump it is usually measured in feet of liquid (refer to Figure 7.4 for a typical method for calculating NPSH available. It is important to note that the pressure at the suction source cannot be considered equal to the NPSH. In Figure 7.4 it can be seen that the source pressure is the same as the vapor pressure, indicating that

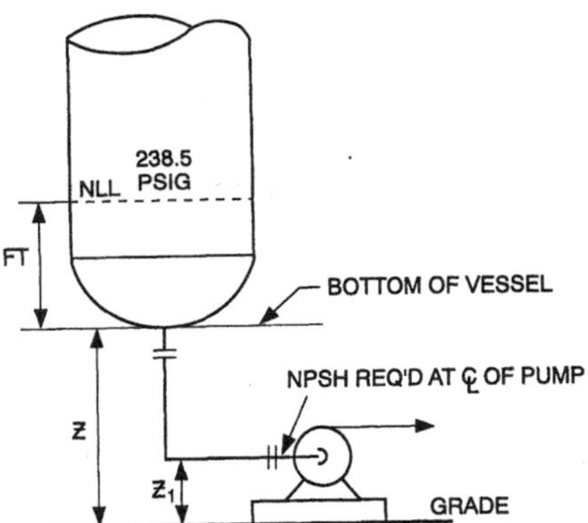

$NPSH_a$ = AVAILABLE NET POSITIVE SUCTION HEAD
P = PRESSURE ABOVE LIQUID INTERFACE, PSIA
P_a = ATMOSPHERIC PRESSURE, PSIA
P_V = VAPOR PRESSURE OF LIQUID, PSIA
h_f = PRESSURE DROP IN SUCTION PIPING DUE TO FRICTION
Z = ELEVATION FROM BOTTOM OF VESSEL TO GRADE
Z_1 = DISTANCE FROM GRADE TO ℄ OF PUMP
S.G. = SPECIFIC GRAVITY
NLL = NORMAL LIQUID LEVEL
S.M. = MARGIN OF SAFETY

$$NPSH_a = \left[\frac{(P + P_a - P_V)2.31}{S.G.}\right] + \left[Z - Z_1 - h_f\right] - S.M.$$

P_V = 253.2 PSIA P_a = 14.7 PSIA Z = 15 FT h_f = 2.4 FT
P = 238.5 PSIG   S.G. = 0.488 Z_1 = 3 FT   SM = 0

$$NPSH_a = \left[\frac{(238.5 + 14.7 - 253.2)2.31}{0.488}\right] + \left[15 - 3 - 2.4\right] - 0 = 9.69$$

SAY 10.

NOTE:
WITH 8FT NLL,
ZERO SAFETY
MARGIN IS ASSUMED

Figure 7.4 Calculate available NPSH

the liquid is at it's boiling point. When the vapour pressure is subtracted from the suction pressure the resulting NPSH available is 2.1 psi or ten (10) feet. When calculating NPSH available it is prudent to incorporate a margin of safety to protect the pump from potential cavitation damage resulting from unexpected upsets. The actual margin amount will vary from company to company. Some will use the normal liquid level as the datum, while others use the vessel tangent or the bottom of the vessel. Typical suggested margins are: two (2) feet for hydrocarbon liquids (including low S.G.), and ten (10) feet for boiling water.

Defining the pump rated point for efficient operation

Since centrifugal pumps are not normally custom designed items of equipment, it is important to assure that each vendor will quote similar pump configurations for the specific operating conditions set forth on each application data sheet. When establishing which pump characteristic and impeller pattern to select for a specific application, certain guidelines should be followed (refer to Figure 7.5).

Application guidelines

- When selecting a specific impeller pattern the rated flow should be no greater than ten (10) percent to the right of best efficiency point. This will result in operation at close to best efficiency point during normal operation (refer to Figure 7.6). Also, selecting a pump to operate too far to the right of best efficiency point can result in the pump operating in the 'break'. A pump is considered operating the 'break' when it is pumping maximum capacity and the total head is reduced while the suction head is held (the impeller actually acts as an orifice to limit the flow).
- Selecting a pump for the rated flow too far to the left of best efficiency point (oversized pump) can result in cavitation damage caused by internal recirculation (refer to Figure 7.7).

Figure 7.5 Application guidelines

Carefully define critical component requirements

Pump reliability improvement can be achieved through proper specification, selection and operation of components such as bearings, mechanical seals and drivers. Industry standards such as API Standard

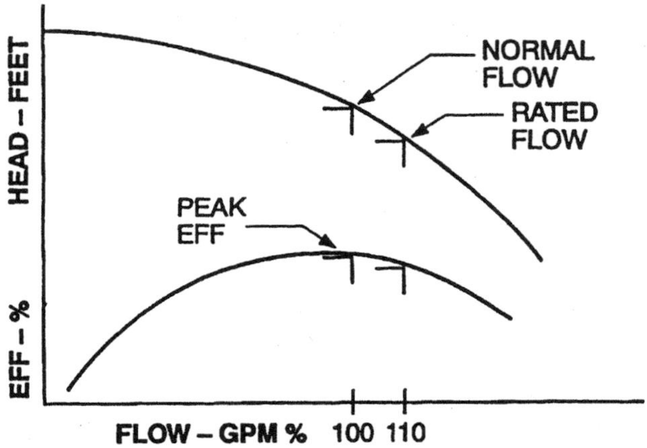

Figure 7.6 Selecting a specific impeller pattern

610 for centrifugal pumps and Standard 682 for mechanical seals contain minimum requirements which, if implemented, should result in improved reliability and extended on stream operating time. Some salient points about each of these components are highlighted in Figures 7.8, 7.9 and 7.10.

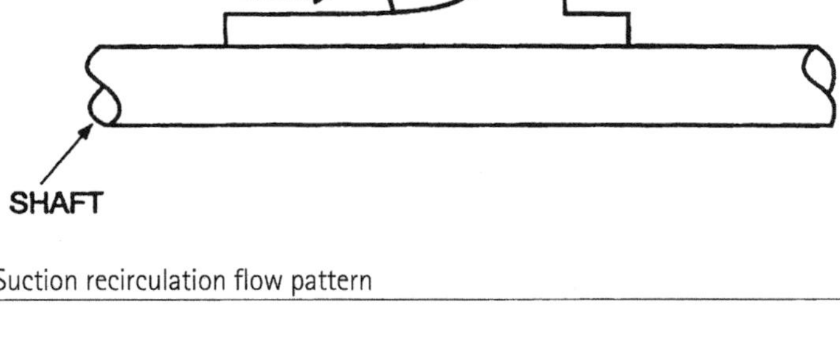

Figure 7.7 Suction recirculation flow pattern

Bearing application guidelines

- Centrifugal pumps require bearings to carry radial/axial loads
- Bearing alternatives: Anti-friction, hydrodynamic ring oil lubricated or hydrodynamic pressure lubricated
- Oil lubricated (anti-friction) bearings are used in majority of process pumps to carry loads
- Pressure lubricated hydrodynamic bearings are normally used for high pressure, high horsepower, high speed applications
- Criteria in API Standard 610 for pressure lubrication: when product of pump rated horsepower and rated speed in rev/min is greater than 2.0 million
- Pressure lubrication systems can either be integral or separate, but should include, as a minimum, an oil pump, reservoir, filter, cooler, controls and instrumentation
- Ring oil lubrication may be applied to hydrodynamic journal bearings in less severe service (when dN factor is less than 300,000. A dN factor is the product of bearing, size (bore) in millimeters and the rated speed in revolutions per minute).

Figure 7.8 Bearing application guidelines

Mechanical seal application guidelines

- Mechanical seals are often used in pumps handling hazardous as well as non-hazardous liquids that must be contained within the unit
- Single seal arrangement is most widely used in process industry
- Single seal design consists of rotary face in contact with a stationary face
- For most services, a carbon face mating against tungsten carbide is satisfactory
- Seals offer the advantages of long life, low maintenance and high reliability
- In general, seals handling light specific gravity liquids at low temperature and high vapor pressure give most problems in the field
- Materials for cold service seals must be suitable for temperatures of start-up, cool down and running; the atmospheric side must be held above 32°F to prevent ice formation; and there must be enough liquid at the seal surfaces

- Successful operation of any seal depends largely on correctly specifying liquid conditions of vapor pressure, temperature specific gravity, etc.
- API Standard 682 is excellent resource for overall mechanical seal application guidelines
- The pressure in the seal chamber (stuffing box) must be at least 25 psig above the pump suction pressure.

Figure 7.9 Mechanical seal application guidelines

Driver sizing guidelines

- Pump drivers are normally electric motor or steam turbine
- Choice of driver type is usually based on plant utility balance plus reliability evaluation of each type to perform within the operating system
- Motors can be sized by several methods:
  - Name plate rating large enough to cover the complete range of pump performance curve
  - Size of motor based on system curve analysis to establish maximum horsepower required at intersection of system curve and pump performance curve
- API Standard 610 has guidelines for sizing motor drives. A margin of 125% is recommended for motors equal to or less than 25hp, 115% for 30 to 75 hp and 110% for motors rated 100hp or more
- Steam turbine drives are normally sized for the power required at pump rated condition. This is possible because turbines can accommodate increased power loads more readily than electric motors.

Figure 7.10 Driver sizing guidelines

Guidelines to use when selecting pump style

The choice for selecting the type of pump to use for a given application can vary with specific gravity, operating temperature, pressure conditions, liquid composition and available NPSH. Figures 7.11 to 7.17 provide guidelines which can help make the choice or selecting a style of pump less complicated.

Single stage, single suction overhung impeller characteristics

- Most commonly applied centrifugal pump - most applications
- Total head limited to 15 inch impeller diameter @3600 rpm (approximately 600 feet head). Larger diameter impellers operate at lower speeds
- Low, medium, high temperature (with cooled bearings, stuffing box)
- Relatively low NPSH required for single suction impeller
- All process services with proper materials selection
- Center of gravity of impeller is outside bearing span
- Axial thrust

Figure 7.11 Single stage, single suction overhung impeller characteristics

Single stage in-line pump characteristics

- Gaining acceptance as alternative to single stage overhung pump
- Total head limited to approximately 400 ft
- Low temperature applications only
- Relatively low NPSH required
- Limited to approximately 200 H.P.
- Most designs utilized do not incorporate bearings (they use the motor bearings to position the pump shaft)
- Note: Designs are now available that incorporate an anti-friction bearing in the pump housing

Figure 7.12 Single stage in-line pump characteristics

Single stage integral gear centrifugal pump characteristics

- Used for high head low flow applications
- Total maximum head approximately 2500 FT.
- Can be used for all temperatures
- Lowest $NPSH_{required}$ (can use inducer)
- Limited to 400 bhp maximum
- All process services with proper material selection
- Incorporates gear box to increase pump speeds as high as 30,000 rpm

Figure 7.13 Single stage integral gear centrifugal pump characteristics

Plunger-P.D. pump characteristics

- Use for high pressure applications (greater than 2000 psi)
- Unlimited head capability
- Maximum flow approximately 500 gpm (5 plungers)
- Limited to approximately 1500 bhp
- Constant volume capacity
- Produces pulsations and requires pulsation dampners

Figure 7.14 Plunger P.D. pump characteristics

Single stage, double suction impeller between bearing pump characteristics

- Used for high flow applications (greater than 2000 gpm)
- Low, medium head requirements
- Low NPSH required
- Low, medium high temperature (with cooled bearings, stuffing box)
- All process services with proper materials selection
- Center of gravity of impeller is between bearing span
- No speed constraint
- Low axial thrust
- Confirm that suction specific speed is less than 9,000. (Definition and equation is in Chapter 8)

Figure 7.15 Single stage, double suction impeller between bearing pump characteristics

Horizontal multistage between bearing pump characteristics

- Used for high head medium flow applications
- Double suction impeller for first stage for low $NPSH_R$
- Low, medium, high temperature (with cooled bearings, stuffing box)
- No speed constraint
- Thrust requires compensation (back to back impellers, balance device)

Figure 7.16 Horizontal multistage between bearing pump characteristics

Vertical multistage pump characteristics

- Used for low NPSH available applications
- High head capability by adding stages
- Low, medium, high temperature
- Low, medium flow range
- No speed constraint
- Most non abrasive process liquids with proper materials selection

Figure 7.17 Vertical multistage pump characteristics

Using the guidelines presented, let us now focus on three examples of how to select a centrifugal pump for a given process system application (refer to Figure 7.18)

Before the appropriate pump and driver can be selected, it will be necessary to completely define the process system operating conditions in which the pump will operate. This will include the suction and discharge system resistance, the head (energy) required by the system and the NPSH available (refer to Figure 7.19).

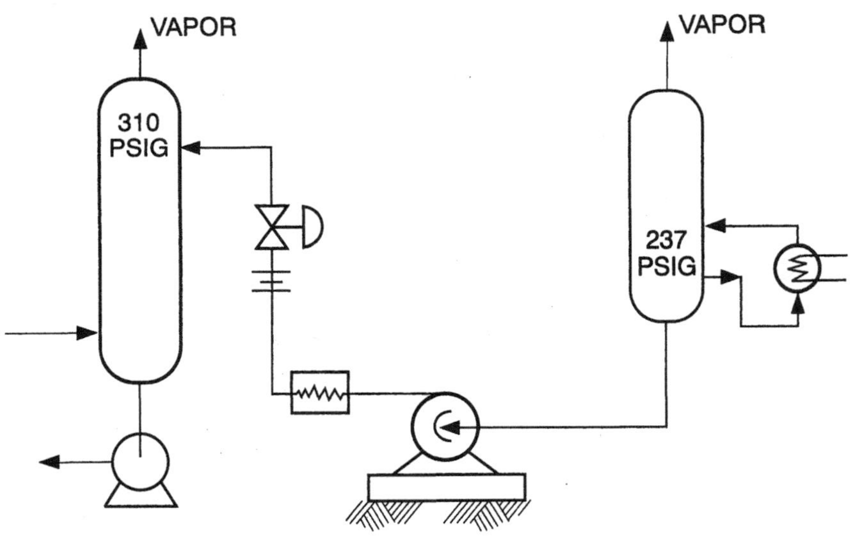

Figure 7.18 Example No. 1 process system

Calculate process system variables

Elevations above grade:

- Pump centerline 3ft
- Inlet nozzle to vessel 72ft
- Liquid surface in suction vessel max 32ft, min 22 ft

Pressure p, in bottom of suction vessel 237 psig

Pressure drops:

- Suction piping 1 psi
- Discharge piping 10 psi
- Flow orifice 2 psi
- Control valve 30 psi
- Exchanger 15 psi

Flow rate: 500 gpm

Specific gravity 0.488 @ p.t.

Vapor pressure 251 psia

Suction pressure calculation:

- Vessel pressure 237 psig
- Static elevation head

(22-3) × 0.433 × 0.488 = 4 psi

- Suction line Δp -1 psi

P_s ($Vessel_{press}$ + $Elev_{press}$ – loss) = 240 psi

Discharge pressure calculation:

- Vessel pressure 310 psig
- Static elevation head

(72-3) × .433 × .488 = 14.5 psi

- Friction ΔP:
- Piping 10 psi
- Orifice 2 psi
- Control valve 30 psi
- Exchanger 15 psi

Pd ($Vessel_{press}$ + $Elev_{press}$ + losses) = 381.5 psi

ΔP Pd –Ps

381.5–240

141.5 psi

$$\frac{\Delta P \times 2.31}{s.g.} = \frac{141.5 \times 2.31}{0.488} = 670 \text{ ft}$$

$NPSH_{AVAILABLE}$ (for boiling liquid)

p_a surface + 14.7 – P_V + static elev diff – friction

237 + 14.7 – 251 + 4 – 1 = 3.7 psi

$$NPSH_{AVAILABLE} = \frac{3.7 \times 2.31}{0.486} = 17.5 \text{ ft}$$

Figure 7.19 Calculate process system variables

When the process system is defined, the next step is to complete the tasks presented in Figure 7.20.

Tasks for selecting pump and driver

- Select pump type based on guidelines:
  - Single stage overhung impeller
  - Multi stage axial or radial split casing design
  - Match NPSHR vs NPSHA
  - Calculate bhp based on pump efficiency
  - Determine driver hp rating based on API criteria

Figure 7.20 Tasks for selecting pump and driver

Based on an assessment of the process system requirements in Figure 7.19 and the guidelines for selecting pump and driver we can determine that the pump defined in Figure 7.21 satisfies all of the guidelines.

For example No. 2 let us select a pump for a boiler feed water application with operating conditions shown in Figure 7.22.

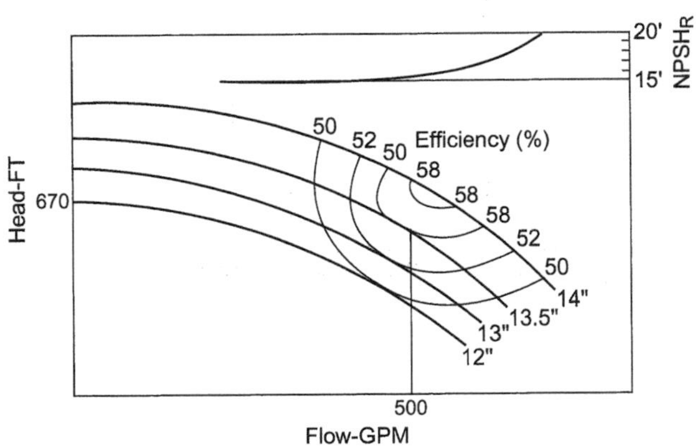

PUMP TYPE 3 x 4 x 14A – HHS

SINGLE STAGE OVERHUNG DESIGN

EFFICIENCY = 56%

$$\text{BHP} = \frac{500 \times 670 \times .488}{3960 \times .56} = 73.7$$

NPSHR = 16 FT

NPSHA = 17.5 FT

SPEED = 3550 RPM

IMPELLER DIA = 13 1/2 INCH

Figure 7.21 Example No 1 pump selection

Example No. 2 operating conditions	
Liquid	**Boiler feed water**
S.G.	0.93
P.T.	220°F
P_s	25 psig
P_d	650 psig
$NPSA_{available}$	26 ft
Flow rate rated	275 gpm
$head_{required}$	1553 ft.

Figure 7.22 Example No. 2 operating conditions

This application requires a multi-stage axial split case pump based on the criteria that the head (energy) required by the system exceeds the head (energy) which can be provided by a 15 inch single stage impeller.

The pump selected is a Union Pump 3 × 4 MOC, 5 stage axial split casing unit (refer to Figures 7.23 and 7.24 for the performance curve and cross section drawing, respectively). Note that selecting a 9.50″ diameter will result in the pump operating at its best efficiency point (bep) at rated flow

Example No. 3 operating conditions	
■ Flow rate	100 gpm
■ S.G.	0.98
■ P.T.	120°F
■ NPSHA	0 ft. at grade
■ P_s	1.96 psia
■ P_d	120 psig
■ Head	3.13

Figure 7.23 Example No. 3 operating conditions

For our third example, we shall examine the selection of a pump type with a constraint on NPSH available. A hot well condensate pump installed in a steam turbine condenser system will be used to illustrate this example (refer to Figure 7.23) for operating conditions.

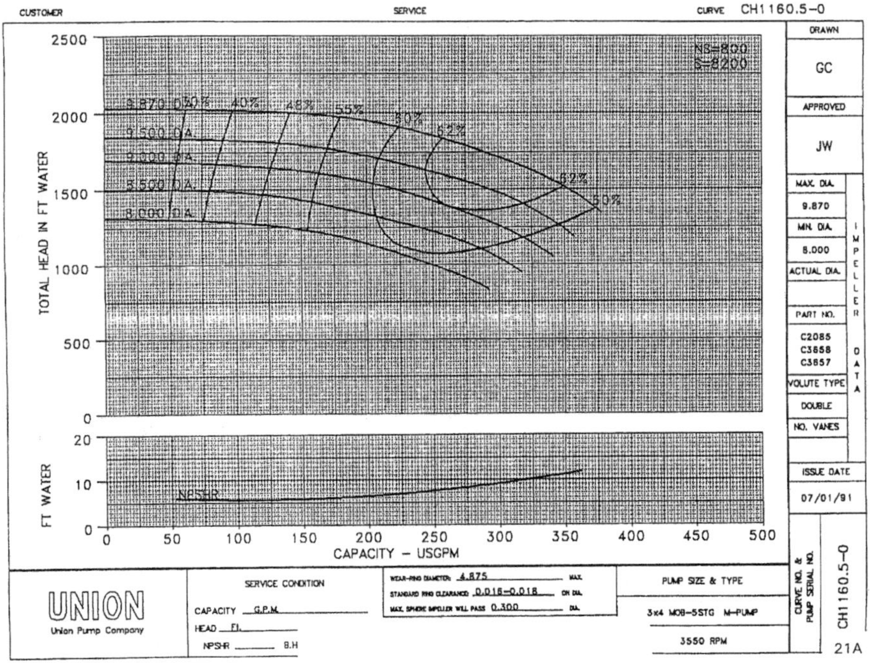

Figure 7.24 (Courtesy of Union Pump Co)

It is apparent that the NPSH available is a major constraint for selecting a conventional horizontal pump for pumping condensate from the condenser hot well. For this application, a vertical canned pump is the appropriate selection (refer to Figure 7.26).

The feature about this design which makes it suitable for use in this type of service is the fact that the first stage impeller is located at the bottom end of the shaft and the shaft length can be made sufficiently

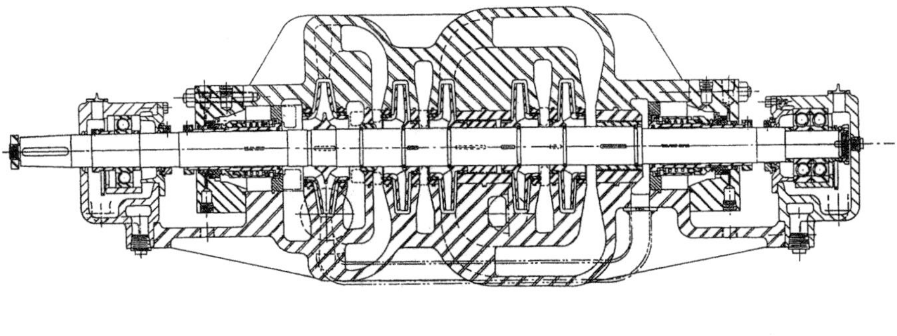

Figure 7.25 Multistage centrifugal pump (Courtesy of Union Pump Co)

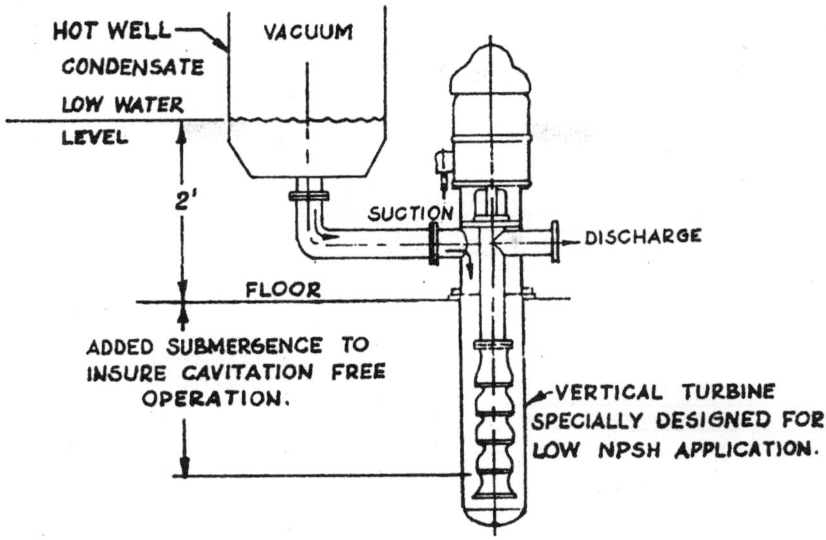

Figure 7.26 Application of vertical pump in condensate hot well service

long to satisfy the NPSH required by the pump. It is common practice to reference available NPSH to grade elevation for this type of pump design. This allows for variations in design of concrete foundation height and location of suction nozzle centerline from top of foundation (refer to Figure 7.27).

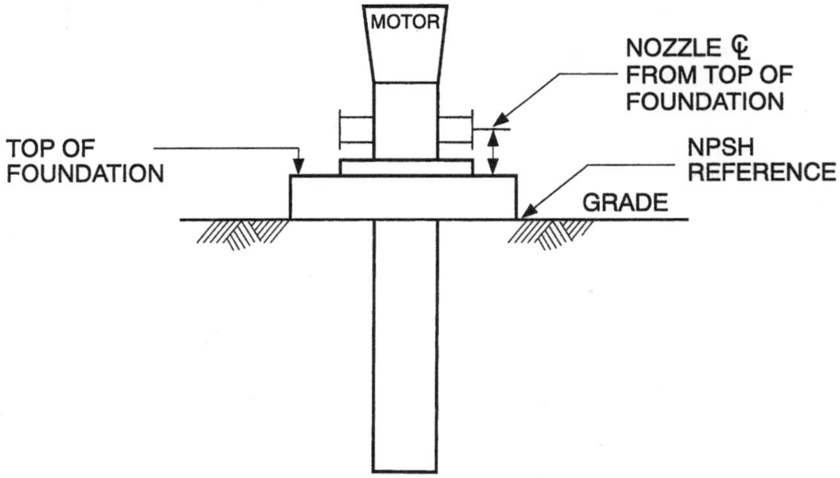

Figure 7.27 Example No. 3 NPSH reference

PAGE 1 OF 5

CENTRIFUGAL PUMP DATA SHEET
CUSTOMARY UNITS

JOB NO. ____ ITEM NO. ____
PURCH. ORDER NO. ____ DATE ____
INQUIRY NO. ____ BY ____
REVISION ____ DATE ____

APPLICABLE TO: ○ PROPOSAL ○ PURCHASE ○ AS BUILT
FOR ____ UNIT ____
SITE ____ NO. REQUIRED ____
SERVICE ____ PUMP SIZE, TYPE & NO. STAGES ____
MANUFACTURER ____ MODEL ____ SERIAL NO. ____

NOTE: ○ INDICATES INFORMATION COMPLETED BY PURCHASER □ BY MANUFACTURER ⊡ BY MANUFACTURER OR PURCHASER

GENERAL

PUMPS TO OPERATE IN (PARALLEL) (SERIES) WITH ____ | NO. MOTOR DRIVEN ____ | NO. TURBINE DRIVEN ____
PUMP ITEM NO. ____ | PUMP ITEM NO. ____
GEAR ITEM NO. 1 ____ | MOTOR ITEM NO. ____ | TURBINE ITEM NO. ____
GEAR PROVIDED BY ____ | MOTOR PROVIDED BY ____ | TURBINE PROVIDED BY ____
GEAR MOUNTED BY ____ | MOTOR MOUNTED BY ____ | TURBINE MOUNTED BY ____
GEAR DATA SHEET NO'S ____ | DRIVER DATA SHEET NO.'S ____ | TURBINE DATA SHEET NO'S ____

○ OPERATING CONDITIONS

○ CAPACITY, NORMAL ____ (GPM) RATED ____ (GPM)
OTHER ____
○ SUCTION PRESSURE MAX/RATED ____ / ____ PSIG
○ DISCHARGE PRESSURE ____ (PSIG)
○ DIFFERENTIAL PRESSURE ____ (PSI)
○ DIFFERENTIAL HEAD ____ (FT) NPSH AVAILABLE ____ (FT)
○ HYDRAULIC POWER ____ (HP)
SERVICE: ○ CONTINUOUS ○ INTERMITTANT (STARTS/DAY ____)

○ SITE AND UTILITY DATA

LOCATION:
○ INDOOR ○ HEATED ○ UNDER ROOF
○ OUTDOOR ○ UNHEATED ○ PARTIAL SIDES
○ GRADE ○ MEZZANINE ○ ____
○ ELECTRIC AREA CLASSIFICATION CL ____ GR ____ DIV ____
○ WINTERIZATION REQD. ○ TROPICALIZATION REQD.

SITE DATA:
○ ELEVATION ____ FT BAROMETER ____ (PSIA)
○ RANGE OF AMBIENT TEMPS: MIN/MAX ____ / ____ °F
○ RELATIVE HUMIDITY: % MAX/MIN ____ / ____

UNUSUAL CONDITIONS: ○ DUST ○ FUMES
○ OTHER ____
○ UTILITY CONDITIONS:

STEAM:

	DRIVERS		HEATING	
MIN	____ PSIG	____ °F	____ PSIG	____ °F
MAX	____ PSIG	____ °F	____ PSIG	____ °F

ELECTRICITY:

	DRIVERS	HEATING	CONTROL	SHUTDOWN
VOLTAGE	____	____	____	____
HERTZ	____	____	____	____
PHASE	____	____	____	____

COOLING WATER:
TEMP. INLET ____ °F MAX RETURN ____ °F
PRESS NORM ____ (PSIG) DESIGN ____ (PSIG)

○ SITE AND UTILITY DATA (CONT'D)

COOLING WATER:
MIN RETURN ____ PSIG MAX ALLOW Δ P ____ (PSI)
WATER SOURCE ____
INSTRUMENT AIR: MAX/MIN PRESS ____ / ____ (PSIG)

○ LIQUID

○ TYPE OR NAME OF LIQUID ____
○ PUMPING TEMPERATURE
NORMAL ____ °F MAX ____ °F MIN ____ °F
○ SPECIFIC GRAVITY ____ @ MAX TEMP
○ SPECIFIC HEAT ____ Cp (BTU/LB °F)
○ VISCOSITY ____ (cP) @ ____ °F
○ MAX. VISCOSITY @ MIN. TEMP. ____ (cP)
○ CORROSIVE/EROSIVE AGENT ____
○ CHLORIDE CONCENTRATION (PPM) ____
○ H_2S CONCENTRATION (PPM) ____
LIQUIDS: (3.5.2.11) ○ TOXIC ○ FLAMMABLE ○ OTHER

□ PERFORMANCE

□ RPM ____
PROPOSAL CURVE NO. ____
□ IMPELLER DIA RATED ____ MAX ____ MIN ____ (IN)
□ RATED POWER ____ (BHP) EFFICIENCY ____ %
□ MINIMUM CONTINUOUS FLOW:
THERMAL ____ (GPM) STABLE ____ (GPM)
□ MAX HEAD RATED IMPELLER ____ (FT)
□ MAX POWER RATED IMPELLER ____ (BHP)
□ NPSH REQUIRED AT RATED CAP. ____ (FT H_2O)
□ SUCTION SPECIFIC SPEED ____
□ MAX SOUND PRESSURE LEVEL ____ dBA
REMARKS: ____

3L/B5390?00835
Printed In U.S.A.
25
DS-610 - 1 11/88

Centrifugal pump data sheet – A

PAGE 2 OF 5

CENTRIFUGAL PUMP DATA SHEET CUSTOMARY UNITS

JOB NO. ______ ITEM NO. ______
REVISION ______ DATE ______
BY ______

☐ CONSTRUCTION

☐ MAIN CONNECTIONS:

	SIZE	ANSI RATING	FACING	POSITION
SUCTION				
DISCHARGE				
BAL. DRUM				

☐ OTHER CONNECTIONS

SERVICE	NO.	SIZE	TYPE
DRAIN			
VENT			
PRESSURE GAUGE			
TEMP GAUGE			
WARM-UP			

CASING MOUNTING:
☐ CENTERLINE ☐ NEAR CENTERLINE
☐ FOOT ☐ SEPARATE MOUNTING PLATE
☐ VERTICAL ☐ SUMP
☐ IN-LINE

CASING SPLIT:
☐ AXIAL ☐ RADIAL

CASING TYPE:
☐ SINGLE VOLUTE ☐ DOUBLE VOLUTE
☐ BARREL ☐ DIFFUSER
☐ STAGGERED VOLUTES ☐ VERTICAL DOUBLE CASING

IMPELLER MOUNTED:
☐ BETWEEN BEARINGS ☐ OVERHUNG
○ IMPELLERS INDIVIDUALLY SECURED (2.5.4)

CASE PRESSURE RATING:
○ SUCTION PRESS. REGIONS OF MULTISTAGE OR DOUBLE CASING PUMP DESIGNED FOR MAXIMUM ALLOWABLE WORK PRESSURE.

ROTATION: (VIEWED FROM COUPLING END)
☐ CW ☐ CCW

REMARKS: ______

SHAFT:
SHAFT DIAMETER AT SLEEVE ______ IN.
SHAFT DIAMETER AT COUPLING ______ IN.
SHAFT DIAMETER BETWEEN BRGS. ______ IN.
SPAN BETWEEN BEARINGS ℄ ______ IN.
SPAN BETWEEN BEARING & IMPELLER ______ IN.
REMARKS: ______

COUPLINGS: DRIVER-PUMP
○ MAKE ______
☐ MODEL ______
☐ CPLG. RATING (HP/100 RPM) ______

COUPLINGS: (CONTINUED)
○ LUBRICATION ______
◙ LIMITED END FLOAT REQUIRED ______
◙ SPACER LENGTH ______ IN.
◙ SERVICE FACTOR ______
○ DYNAMIC BALANCED AGMA BALANCE CLASS ______
DRIVER HALF COUPLING MOUNTED BY
○ PUMP MFR. ○ DRIVER MFR. ○ PURCHASER
○ COUPLING PER API 671
REMARKS: ______

MATERIAL

○ TABLE H-1 CLASS ______
☐ BARREL/CASE ______ IMPELLER ______
☐ CASE/IMPELLER WEAR RINGS ______
☐ SHAFT ______ SLEEVE ______
☐ DIFFUSERS ______
☐ COUPLING HUBS ______
☐ COUPLING SPACER ______
☐ COUPLING DIAPHRAGMS ______
☐ API BASEPLATE NUMBER / MATERIAL ______ / ______
○ VERTICAL LEVELING SCREWS (3.3.1.15) ______
○ HORIZONTAL POSITIONING SCREWS (3.3.1.14) ______
REMARKS ______

BEARINGS AND LUBRICATION

BEARING: (TYPE / NUMBER)
☐ RADIAL ______ / ______
☐ THRUST ______ / ______
○ REVIEW AND APPROVE THRUST BEARING SIZE

LUBRICATION
☐ GREASE ☐ FLOOD ○ RING OIL
☐ FLINGER ☐ PURGE OIL MIST ○ PURE OIL MIST
○ CONSTANT LEVEL OILER
○ PRESSURE ○ API-610 ○ API-614
○ OIL VISC, ISO GRADE ______
◙ OIL HEATER REQ'D ○ ELECTRIC ○ STEAM
○ OIL PRESSURE TO BE GREATER THAN COOLANT PRESSURE
REMARKS ______

3L/B5390?00835
Printed in U.S.A.

26

DS-610 - 2 11/88

Centrifugal pump data sheet – B

PAGE 3 OF 5

CENTRIFUGAL PUMP DATA SHEET
CUSTOMARY UNITS

JOB NO. ______ ITEM NO. ______
REVISION ______ DATE ______
BY ______

MECHANICAL SEAL OR PACKING

SEAL DATA:

○ SPECIAL SEAL DATA SHEET ______
○ API MATERIAL CLASS CODE (TABLE H-4) ______
☒ SEAL MANUFACTURER ______
☒ SIZE AND TYPE ______
☒ MANUFACTURER CODE ______

SEAL CONSTRUCTION:

○ CARTRIDGE MOUNT
○ HOOKED SLEEVE OR NON-CARTRIDGE ○ NO SLEEVE
○ PUMPING RING
☐ SLEEVE MATERIAL ______
☐ GLAND MATERIAL ______
○ AUX. SEAL DEVICE ______
☒ JACKET REQUIRED

GLAND TAPS:

☒ FLUSH (F) ☒ DRAIN (D) ☒ BARRIER (B)
☒ VENT (V) ☒ COOLING (C)
☒ QUENCH (Q) ☒ HEATING (H)

SEAL FLUIDS REQUIREMENT AND AVAILABLE FLUSH LIQUID:

NOTE: IF FLUSH LIQUID IS PUMPAGE LIQUID (AS IN FLUSH PIPING) PLANS 11 TO 41), FOLLOWING FLUSH LIQUID DATA IS NOT REQUIRED.

○ TEMPERATURE (SUPPLY) ______ °F
○ TEMPERATURE MIN / MAX ______ / ______ °F
○ SPECIFIC GRAVITY ______ @ ______ °F
○ NAME OF FLUID ______
○ SPECIFIC HEAT ______ Cp(BTU/LB°F)
○ VAPOR PRESSURE ______ PSIA @ ______ °F
○ TOXIC ○ FLAMMABLE ○ OTHER ______
☐ FLOW RATE MAX / MIN ______ / ______ (GPM)
☐ PRESSURE REQUIRED MAX / MIN ______ / ______ (PSIG)
☐ TEMPERATURE REQUIRED MAX / MIN ______ / ______ °F

BARRIER FLUID:

○ SUPPLY TEMPERATURE MIN / MAX ______ / ______ °F
○ SPECIFIC GRAVITY ______ @ ______ °F
○ NAME OF FLUID ______
○ VAPOR PRESSURE ______ (PSIA) @ ______ °F
○ TOXIC ○ FLAMMABLE ○ OTHER ______
☐ FLOW RATE MAX / MIN ______ / ______ (GPM)
☐ PRESSURE REQUIRED MAX/MIN ______ / ______ (PSIG)
☐ TEMPERATURE REQUIRED MAX / MIN ______ / ______ °F

QUENCH FLUID:

○ NAME OF FLUID ______
○ FLOW RATE ______ GPM

SEAL FLUSH PIPING:

○ SEAL FLUSH PIPING PLAN ______
○ TUBING ○ CARBON STEEL
○ PIPE ○ STAINLESS STEEL

SEAL FLUSH PIPING: (CONT'D.)

○ AUXILIARY FLUSH PLAN ______
○ TUBING ○ CARBON STEEL
○ PIPE ○ STAINLESS STEEL
○ PIPING ASSEMBLY:
○ THREADED ○ UNIONS ○ SOCKET WELDED
○ SEAL WELDED ○ FLANGED
○ TYPE TUBE FITTINGS ______
○ FLOW INDICATOR (PLAN 52/53)
○ PRESSURE SWITCH (PLAN 52/53)
○ PRESSURE GAUGE (PLAN 52/53)
○ TEMPERATURE INDICATOR (PLANS 21, 22, 23, 32, 41)
○ HEAT EXCHANGER (PLANS 52/53)

PACKING DATA:

MANUFACTURER ______
TYPE ______
SIZE. AND NO. RINGS ______
☐ PACKING INJECTION REQUIRED
☐ FLOW ______ (GPM) @ ______ (PSIG)
☐ LANTERN RING ______

COOLING WATER PIPING

○ SIGHT FLOW INDICATORS
○ MANIFOLD OUTLET VALVE
○ GALVANIZED PIPING REQUIRED
○ COPPER TUBING REQUIRED
○ STAINLESS STEEL TUBING REQUIRED
☐ COOLING WATER REQUIREMENTS
SEAL JACKET / PEDESTAL / BRG HSG ______ GPM @ ______ (PSIG)
SEAL HEAT EXCHANGER ______ GPM @ ______ (PSIG)
QUENCH ______ GPM @ ______ (PSIG)
TOTAL COOLING WATER ______ GPM

REMARKS ______

INSTRUMENTATION

VIBRATION:
○ NONCONTACTING (API 670) ○ ACCELEROMETER
○ PROVISION FOR MOUNTING ONLY
○ SEE ATTACHED API-670 DATA SHEET

REMARKS: ______

TEMPERATURE AND PRESSURE:

○ RADIAL BRG. METAL TEMP ○ THRUST BRG. METAL TEMP.
○ PROVISION FOR INSTRUMENTS ONLY
○ SEE ATTACHED API-670 DATA SHEET
○ TEMPERATURE GAUGES ○ THERMOWELLS
OTHER: ______
○ PRESSURE GAUGE TYPE ______
LOCATION ______

3L/B5390700835
Printed in U.S.A.

27

DS-610 - 3 11/88

Centrifugal pump data sheet – C

PAGE 4 OF 5

CENTRIFUGAL PUMP DATA SHEET CUSTOMARY UNITS

JOB NO. ____ ITEM NO. ____
REVISION ____ DATE ____
BY ____

INSTRUMENTATION (CONT'D)

TEMPERATURE AND PRESSURE (CONT'D)

○ PRESSURE SWITCH TYPE ____

LOCATION ____

REMARKS ____

○ SPARE PARTS (TABLE 12)

○ START-UP

○ RECONDITIONING

○ CRITICAL SERVICE

○ SPECIFY ____

MOTOR DRIVE

☐ MANUFACTURER ____

☐ ____ HP ⊡ ____ RPM

☐ HORIZONTAL ☐ VERTICAL

☐ FRAME ____

☐ SERVICE FACTOR ____

○ VOLTS/PHASE/HERTZ ____

○ TYPE ____

☐ ENCLOSURE ____

○ EXPLOSIONPROOF "T" CODE RATING ____

○ MINIMUM STARTING VOLTAGE ____

○ TEMPERATURE RISE ____

⊡ FULL LOAD AMPS ____

⊡ LOCKED ROTOR AMPS ____

⊡ INSULATION ____

⊡ STARTING METHOD ____

⊡ BEARINGS ____

⊡ LUBE ____

○ 2X THRUST RATING (3.1.4)

VERTICAL SHAFT: ○ SOLID ○ HOLLOW

☐ VERTICAL THRUST CAPACITY

UP ____ LBS DOWN ____ LBS

REMARKS ____

VERTICAL PUMPS

☐ PUMP THRUST	UP		DOWN	
AT MIN FLOW	____	(LBS)	____	(LBS)
AT DESIGN FLOW	____	(LBS)	____	(LBS)
AT RUNOUT	____	(LBS)	____	(LBS)
MAX THRUST	____	(LBS) @	____	(GPM)

○ SEPARATE MOUNTING PLATE

○ DRIVE COMPONENT ALIGNMENT SCREWS

○ PIT OR SUMP DEPTH ____ (FT)

☐ PUMP LENGTH ____ (FT)

VERTICAL PUMPS (CONT'D)

☐ MIN. SUBMERGENCE REQUIRED ____ FT

COLUMN PIPE: ☐ FLANGED ☐ THREADED

LINE SHAFT: ☐ OPEN ☐ ENCLOSED

GUIDE BUSHINGS:

☐ BOWL ____ ☐ LINE SHAFT ____

GUIDE BUSHINGS LUBE:

☐ WATER ☐ OIL

☐ GREASE ☐ PUMPAGE

REMARKS ____

APPLICABLE SPECIFICATIONS

API 610, CENTRIFUGAL PUMP FOR GEN. REFINERY SERV.

○ VENDOR HAVING UNIT RESPONSIBILITY ____

○ GOVERNING SPECIFICATION (IF DIFFERENT) ____

REMARKS ____

SURFACE PREPARATION AND PAINT

○ MANUFACTURER'S STANDARD:

○ OTHER ____

PUMP:

○ PUMP SURFACE PREPARATION SSPC-SP- ____

○ PRIMER ____

○ FINISH COAT ____

BASEPLATE

○ BASEPLATE SURFACE PREPARATION SSPC-SP- ____

○ PRIMER ____

○ FINISH COAT ____

GROUTING REQ'D. (3.3.1.17) ○ YES ○ NO

○ GROUT SURFACE PREPARATION SSPC-SP- ____

○ EPOXY PRIMER ____

○ REMARKS ____

SHIPMENT: (4.4.1)

○ DOMESTIC ○ EXPORT ○ EXPORT BOXING REQD.

○ OUTDOOR STORAGE MORE THAN 6 MONTHS

SPARE ROTOR ASSEMBLY PACKAGED FOR :

○ HORIZONTAL STORAGE ○ VERTICAL STORAGE

○ TYPE OF SHIPPING PREPARATION ____

REMARKS: ____

☐ WEIGHTS

MOTOR DRIVEN:

WEIGHT OF PUMP (LBS.) ____

WEIGHT OF BASEPLATE (LBS.) ____

WEIGHT OF MOTOR (LBS.) ____

28

3L/B5390?00835
Printed in U.S.A.

DS-610 - 4 11/88

Centrifugal pump data sheet – D

PAGE 5 OF 5

CENTRIFUGAL PUMP DATA SHEET
CUSTOMARY UNITS

JOB NO. ______ ITEM NO. ______
REVISION ______ DATE ______
BY ______

☐ WEIGHTS (CONT'D)

MOTOR DRIVEN (CONT'D):
- WEIGHT OF GEAR (LBS.) ______
- TOTAL WEIGHT (LBS.) ______

TURBINE DRIVEN:
- WEIGHT OF BASEPLATE (LBS.) ______
- WEIGHT OF TURBINE (LBS.) ______
- WEIGHT OF GEAR (LBS.) ______
- TOTAL WEIGHT (LBS.) ______

REMARKS: ______

○ OTHER PURCHASER REQUIRMENTS

- ○ COORDINATION MEETING REQUIRED ______
- ○ REVIEW FOUNDATION DRAWINGS
- ○ REVIEW PIPING DRAWINGS
- ○ OBSERVE PIPING CHECKS
- ○ OBSERVE INITIAL ALIGNMENT CHECK
- ○ CHECK ALIGNMENT AT OPERATING TEMP.
- ○ ROTOR BALANCED DURING ASSEMBLY OF EACH ELEMENT
- ○ VENDOR DEMONSTRATION OF MAX ALLOW. VIBRATION AT MINIMUM FLOW
- ○ LATERAL RESPONSE ANALYSIS REQUIRED
  - ○ PUMP ONLY ○ ALL EQUIPMENT
- ○ CRITICAL SPEED ANALYSIS
- ○ STIFFNESS MAP OF UNDAMPED ROTOR
- ○ TORSIONAL ANALYSIS
- ○ PROGRESS REPORTS REQUIRED

REMARKS: ______

QA INSPECTION AND TEST

- ○ REVIEW VENDORS QA PROGRAM ______
- ○ PERFORMANCE CURVE APPROVAL
- ○ SHOP INSPECTION

TEST	NON-WIT	WIT	OBSERVE
HYDROSTATIC	○	○	○
PERFORMANCE	○	○	○
NPSH	○	○	○
COMPLETE UNIT TEST	○	○	○
SOUND LEVEL TEST	○	○	○
○ DISMANTLE & INSP. AFTER TEST	○	○	○
○ CLEANLINESS PRIOR TO FINAL ASSEMBLY	○	○	○
○ PIPELOAD TEST	○	○	○
○ ______	○	○	○
○ ______	○	○	○
○ ______	○	○	○
○ ______	○	○	○

QA INSPECTION AND TEST (CONT'D)

- ○ MATERIAL CERTIFICATION REQUIRED
  - ○ CASING ○ IMPELLER ○ SHAFT
  - ○ OTHER ______
- ○ CASTING REPAIR PROCEDURE APPROVAL REQ'D
- ○ INSPECTION REQUIRED FOR NOZZLE WELDS
  - ○ MAG. PARTICLE ○ LIQ. PENETRANT
  - ○ RADIOGRAPHIC ○ ULTRASONIC
- ○ INSPECTION REQUIRED FOR CASTINGS
  - ○ MAG. PARTICLE ○ LIQ. PENETRANT
  - ○ RADIOGRAPHIC ○ ULTRASONIC
- ○ CHARPY IMPACT TEST REQUIRED FOR ______
- ○ INSPECTION REQ'D/ FOR ______
  - ○ MAG. PARTICLE ○ LIQ. PENETRANT
  - ○ RADIOGRAPHIC ○ ULTRASONIC
- ○ HARDNESS TEST REQ'D FOR ______
- ○ SURFACTANT HYDROTEST (4.3.2.5)
- ○ VENDOR SUBMIT TEST PROCEDURES
- ○ RECORD FINAL ASSEMBLY RUNNING CLEARANCES
- ○ VENDOR MAINT. RECORDS FOR 5 YEARS

REMARKS: ______

3L/B5390?00835
Printed in U.S.A.

29

DS-610 - 5 11/86

Centrifugal pump data sheet – E

8 Centrifugal pump hydraulic disturbances

- Introduction
- Maintaining a liquid inside a pump
- Causes of damage
- Preventing hydraulic disturbances
- The project design phase
- Field operation

Introduction

Liquid disturbances in centrifugal pumps are the major cause of reduction in pump reliability. The pressures generated by cavitation can exceed 100,000 PSI. Cavitation caused by many different factors is responsible for pump lost service time as a result of various pump component failures shown in Figure 8.1

Cavitation caused pump component failures

- Bearings
- Seals
- Impellers
- Shaft
- Wear rings

Figure 8.1 Cavitation caused pump component failures

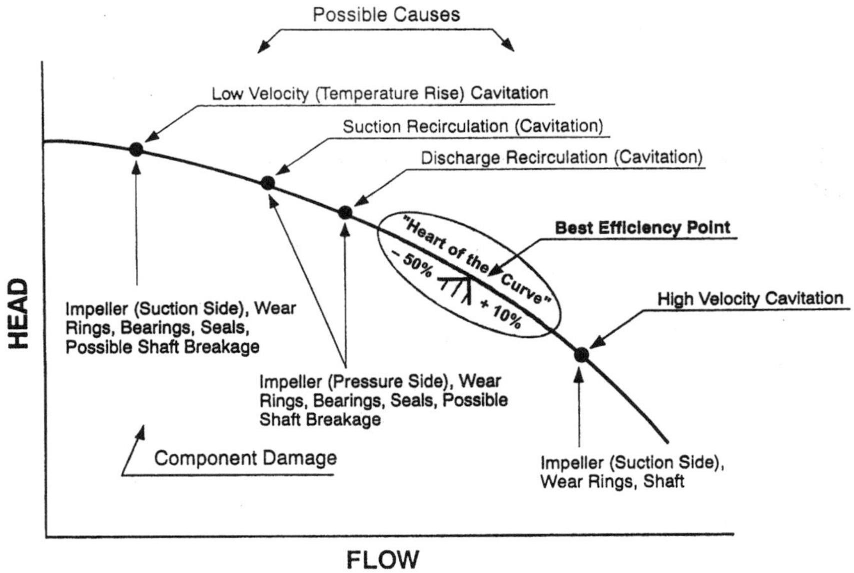

Figure 8.2 Centrifugal pump component damage and causes as a function of operating point

In the previous chapter we discussed proper pump selection guidelines that require a centrifugal pump to be selected in the 'heart of the curve' and that sufficient NPSH available be present to avoid mechanical damage (See Figure 8.2). The objective of the this chapter is to explain the reason for these requirements in detail and provide useful guidelines for preventing liquid disturbances in centrifugal pumps and for solving persistent, costly field pump problems caused by liquid disturbances.

Maintaining a liquid inside a pump

Design objectives

All pumps are designed to increase the energy of a fluid while, maintaining the fluid in its liquid state. Each centrifugal pump is designed to produce a specific amount of head at a specific flow rate based on a specified fluid density. Once the pump is designed, any reduction in the fluid density will result in a reduction in flow rate since the pump now requires greater head (energy) and can produce this value only at a reduced flow rate (See Figure 8.3).

One way to rapidly reduce the density of a fluid is to change its phase. If a liquid suddenly changes to a vapor, the impeller head required to meet the same process differential pressure requirements increases by a factor of 300 or more (See Figure 8.4). As a result, the pump will not be able to move the product since the maximum head produced by the

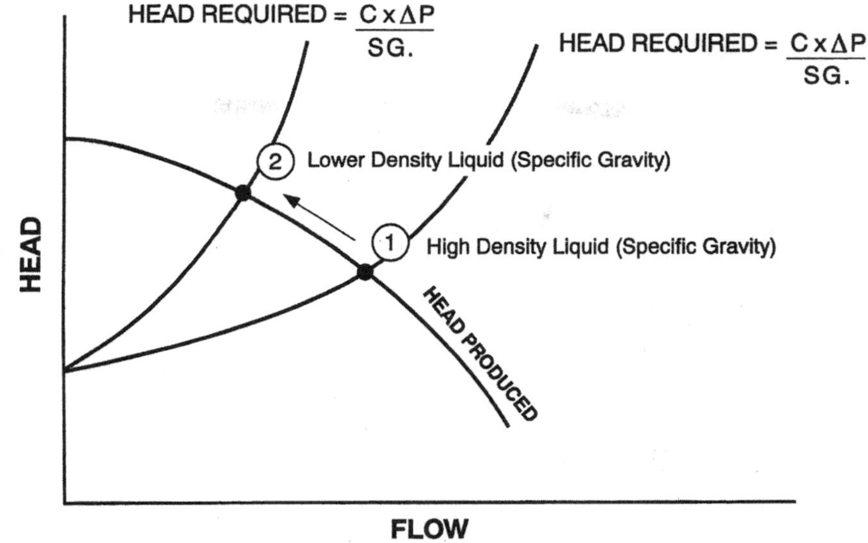

Figure 8.3 Head required by lower density fluid

pump will be much less than the head now required. The discharge pressure gauge will drop in pressure. This is commonly known as 'vapor lock' or 'loss of suction'.

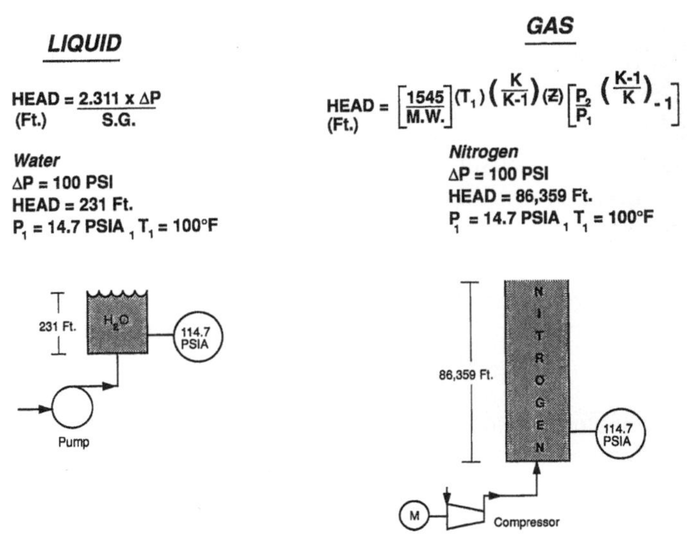

Figure 8.4 Fluid head

Vapor pressure

How can a liquid change phase inside a pump? The vapor pressure for any fluid is that pressure at which the fluid changes from its liquid to vapor phase. Vapor pressure changes with fluid temperature. The vapor pressure for any fluid can be obtained from its Mollier diagram. For a given fluid temperature, the vapor pressure is the intersection of that temperature and the saturated liquid line. Figure 8.5 is the Mollier diagram for ethylene.

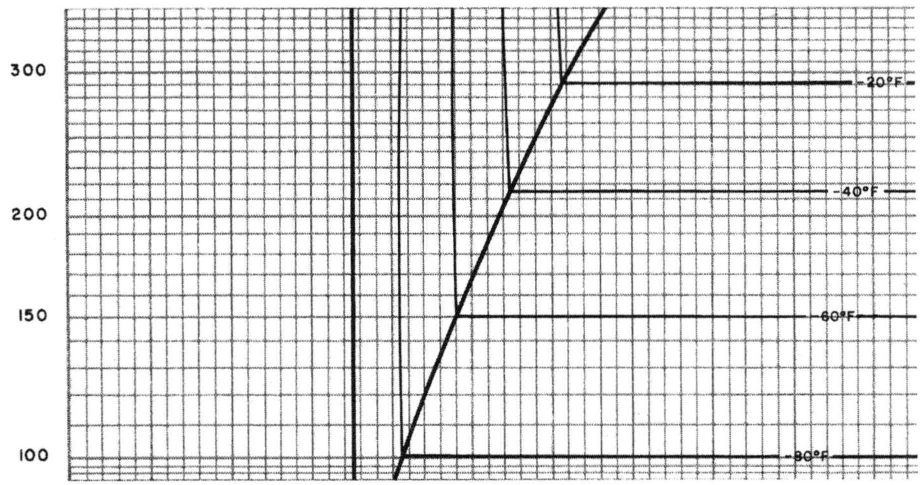

Figure 8.5 Ethelyne Mollier diagram

As can be seen from this figure, either a reduction in pressure (holding fluid temperature constant) or an increase in temperature (holding pressure constant) will cause a phase change. Another way of stating these facts is presented in Figure 8.6.

Maintaining a liquid

A fluid will remain a *liquid* as long as its vapor pressure is less than the pressure acting on the liquid

Figure 8.6 Maintaining a liquid

Before we leave this subject, let's examine how water can change phase (refer to Figure 8.7).

The most common cause of phase change for water is to increase its temperature at a constant atmospheric pressure. When the vapor pressure exceeds the pressure surrounding the liquid (14.7 PSIA) the liquid will change phase or 'boil'. However, water can also change phase 'boil' if its pressure is decreased when temperature is held constant (See Figure 8.7). When the pressure of the water is less than its vapor pressure, water vapor will form. Therefore, any action inside a pump that will either reduce the pressure of a liquid or significantly increase its temperature can cause the liquid to change to a vapor.

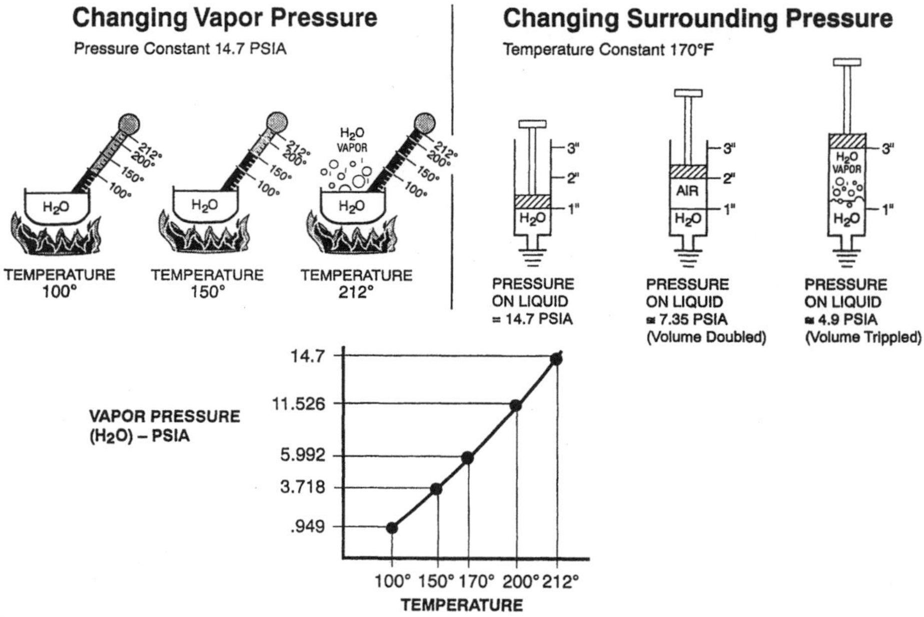

Figure 8.7 Causes of liquid phase change - H_2O

What can reduce the pressure of a liquid or significantly increase its temperature inside a pump? To answer this question, the entire cross section of a pump must be examined. In Figure 8.8, the flow path from the pump suction flange to the impeller vane is plotted vs pressure in the pump. It can be seen that the pressure decreases from the suction to the leading edge of the impeller vanes then increases rapidly once the liquid is inside the impeller vanes where vane tip speed (U) and liquid tangential velocity (V_T) increase head (energy). The total pressure reduction from the suction flange to the impeller vanes is the result of:

- Friction losses in the flow passage
- Liquid acceleration
- Entry shock losses at the impeller vane tips

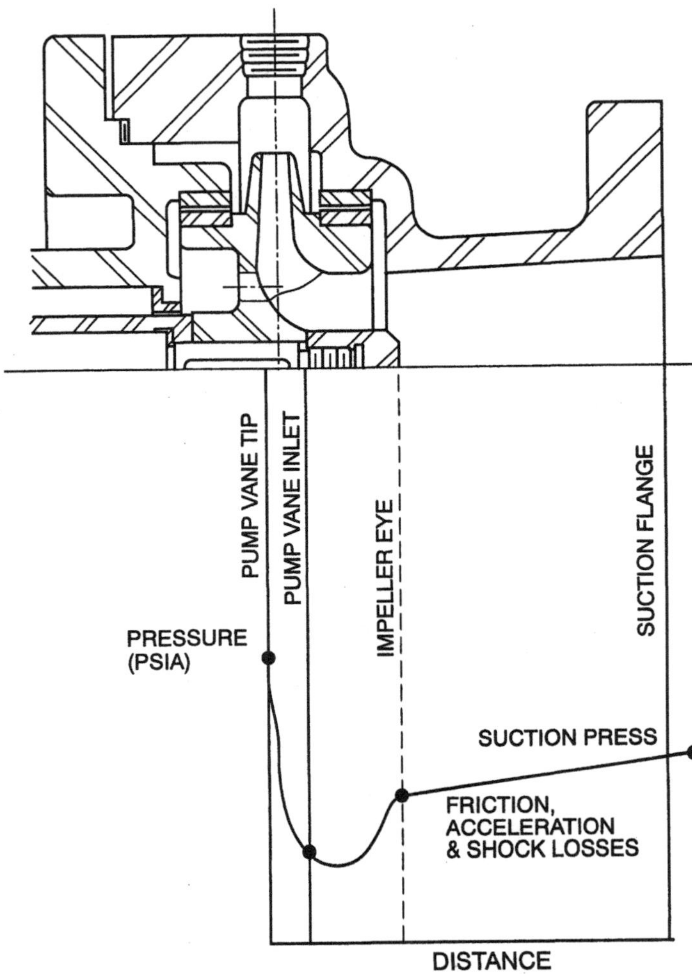

Figure 8.8 Flange to vane entrance losses

Therefore, each pump has a distinct pressure drop for a given flow which is the result of pump case, inlet volute and impeller design. If the pressure drop from the suction flange to the impeller vane reduces the pressure below the liquid's vapor pressure, vapor will be formed at the impeller vanes. There are also two (2) other causes of vapor formation within an impeller.

Low flow velocities at any location within the impeller can cause separation of flow stream lines and lead to low pressure areas (cells) inside the impeller. If the pressure within these cells is less than the fluids' vapor pressure, vapor will form (refer to Figures 8.9 and 8.10).

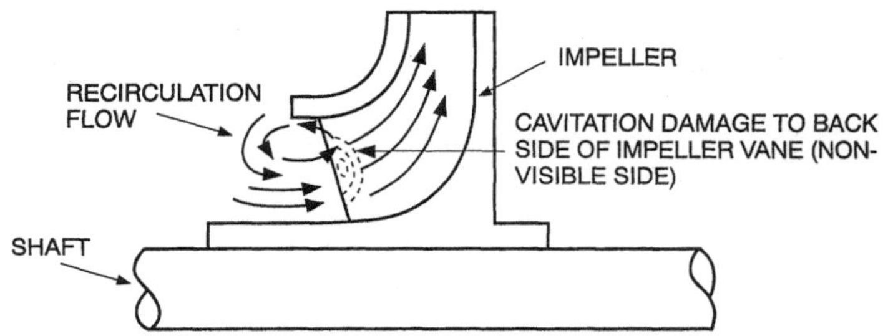

Figure 8.9 Recirculation flow pattern in impeller at low flows

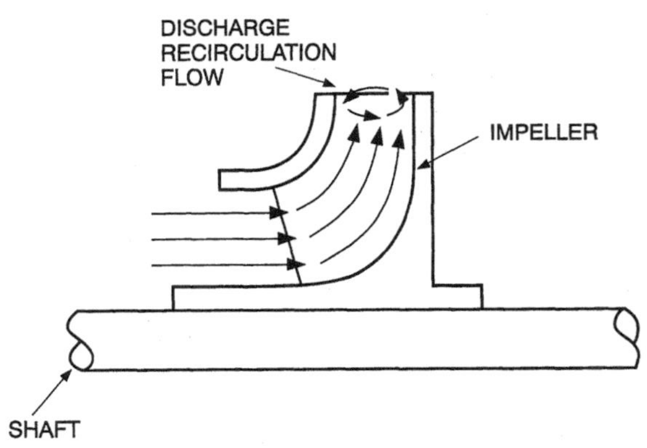

Figure 8.10 Discharge recirculation flow pattern

The curvature of the impeller vanes will always result in lower velocities on the pressure side of the vane (the side that cannot be readily observed). Therefore, vaporization caused by flow separation occurs on the pressure or 'backside' of the vane. This phenomena is commonly known as recirculation.

1. Pump efficiency is low at low flows

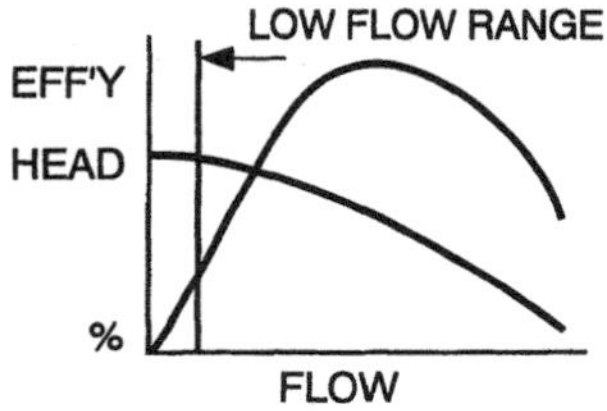

2. Liquid temperature rise increases by

$$\Delta T = \frac{\text{Pump head}}{778 \times Cp} \times \left[\frac{1}{\text{Pump efficiency}} - 1 \right]$$

Where: Pump head is calculated from data in $\frac{FT\text{-}LB_F}{LB_M}$

Cp is specific heat of the fluid in $\frac{BTU}{°F\text{-}LBmass}$

778 is conversion factor $\frac{FT\text{-}LB_F}{BTU}$

Pump efficiency is expressed as a decimal

3. If the temperature rise increases the fluid's vapor pressure above the surrounding pressure, the fluid will vaporize

Note: This action should also be taken for Figure 6.11 in Volume 1

Figure 8.11 Low flow temperature rise can cause vapor formation

The other cause of vapor formation within a pump is the temperature rise associated with low flows (Refer to Figure 8.11).

When operating at low flows, the impeller efficiency reduces significantly, thus increasing the liquid temperature rise. As previously mentioned, the vapor pressure of any liquid increases with temperature. Increased temperature can cause the liquid to vaporize at the impeller vanes. Referring to Figure 8.11, it can be seen that low specific gravity liquids with high vapor pressures are most susceptible to vaporization caused by low flow operation. Note that increased wear ring clearances can worsen this situation since the higher temperature liquid will mix with the cooler liquid entering the impeller. Figure 8.12 summarizes the causes of vaporization within a centrifugal pump.

Causes of vaporization within a centrifugal pump

- Internal inlet pressure losses
- Formation of low pressure cells at low flows
- Liquid temperature rise at low flows

Figure 8.12 Causes of vaporization within a centrifugal pump

Causes of damage

In the above section the causes of vapor formation within a pump were described. In this section the causes of damage to pump components will be discussed.

Cavitation

Figure 8.13 presents the definition of cavitation.

Cavitation definition

Cavitation is the result of released energy when an increase of pressure surrounding the fluid causes the saturated vapor to change back to a liquid

Figure 8.13 Cavitation definition

From the definition of above, vapor must be present before cavitation can take place. The sources of vapor formation were discussed and are summarized in Figure 8.12. Referring back to Figure 8.8, it can be seen that as soon as the liquid enters the impeller vane area, energy and pressure rapidly increase. When the pressure of the liquid exceeds the fluid's vapor pressure, the vapor bubbles will collapse and cavitation will occur. The solution to preventing cavitation is shown in Figure 8.14.

Preventing cavitation

Cavitation is prevented by preventing vapor formation within a pump

Figure 8.14 Preventing cavitation

Methods to prevent cavitation will be discussed later on in this chapter.

The effects of fluids on component damage

The energy released during cavitation caused by inlet pressure losses, recirculation or low flow temperature rise varies as a function of the fluid type and the amount of vaporization. In Figure 8.15, we have drawn a generic Mollier diagram to show that the fluid type (latent heat of vaporization for a given temperature) and degree of vaporization determine the energy released when the vapor is recompressed to a liquid. Note that the abcissa is BTU/LB. When cavitation occurs, a given amount of energy (BTU/LB) will be transferred from the fluid to the impeller. Energy can also be expressed in FT-LB_F/LB_M by multiplying

$$\frac{BTU}{LB_M} \times \frac{778\ FT\text{-}LB_F}{BTU}$$

It can therefore be seen that in cavitation the energy transferred by the fluid to the vanes can greatly exceed the head produced by the impeller. Refer back to Figure 8.5 and determine the maximum amount of energy transferred to the impeller vanes when operating at 20 PSIG.

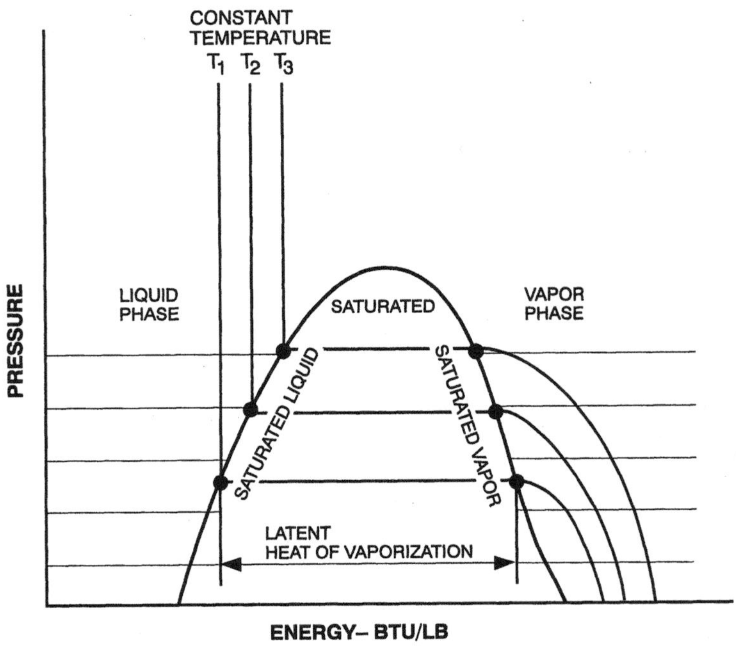

Figure 8.15 Mollier diagram

In general, single component liquids produce the highest energy values during cavitation and are therefore the most damaging fluids. Hydrocarbon mixtures produce lower energy values and have higher viscosities which reduce damage. Regardless of composition, all liquids produce high noise levels during cavitation which typically sounds like solids 'rocks' are passing through the pump. Always remember that there are different causes of vaporization which result in cavitation.

Preventing hydraulic disturbances

In the previous sections, the causes of liquid disturbances in centrifugal pumps were discussed. In this section, practical advice on how to prevent purchasing troublesome pumps in the design phase and practical solutions on how to solve existing field problems caused by liquid disturbances will be presented.

The project design phase

Action taken during the early stages of a project can significantly increase pump reliability and safety by eliminating all sources of vaporization within a pump. Sources of vaporization exist both in the process and in the pump. Before presenting solutions, a number of important concepts must be reviewed and presented.

Concepts

Specific speed

Specific speed is a non-dimensional value that is a function of pump speed, flow and head

$$N_S = \frac{N\sqrt{Q}}{HD^{3/4}}$$

Where: N_S = specific speed
N = pump speed rpm
Q = pump flow gpm
Hd = pump produced head $\frac{FT/LB_F}{LB_M}$

Note: For double suction impellers, Q = Q/2

Specific speed is used extensively in both pump and compressor design to optimize stage efficiency for a given value of flow and head required. In pump design, specific speed is used to optimize the following design parameters:

- Impeller discharge flow velocity
- Impeller tip speed
- Impeller inlet and discharge blade angles
- Discharge throat velocity

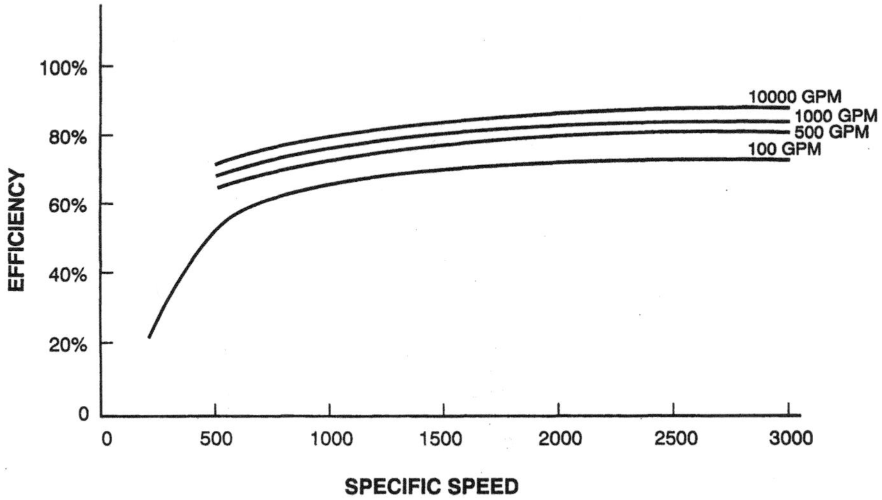

Figure 8.16 Efficiency as a function of pump specific speed and flow

Figure 8.16 presents a plot of pump specific speed vs efficiency for various flow rates. This is a typical chart and will vary slightly from pump vendor to pump vendor based on the specifics of pump design. It can be used for estimating purposes in determining pump efficiency to obtain pump required horsepower.

Net positive suction head available

$NPSH_A$ has been previously discussed. As we have learned, it must be greater than the $NPSH_R$ to prevent cavitation. Methods for determining $NPSH_A$ have been presented.

Net Positive Suction Head Required

Figure 8.17 shows the NPSH required within a typical centrifugal pump. It can be seen that this value is actually the pressure drop from the suction flange to the impeller vane inlet expressed in energy terms $\frac{(FT/LB_F)}{LB_M}$.

Perhaps now, we can truly understand why $NPSH_{AVAILABLE}$ must be $\geq NPSH_{REQUIRED}$. As we have learned,

- *If* $NPSH_A \geq NPSH_R$
- *Then*, the Fluid will not Vaporize
- *Therefore*, no vaporization, no cavitation, no damage

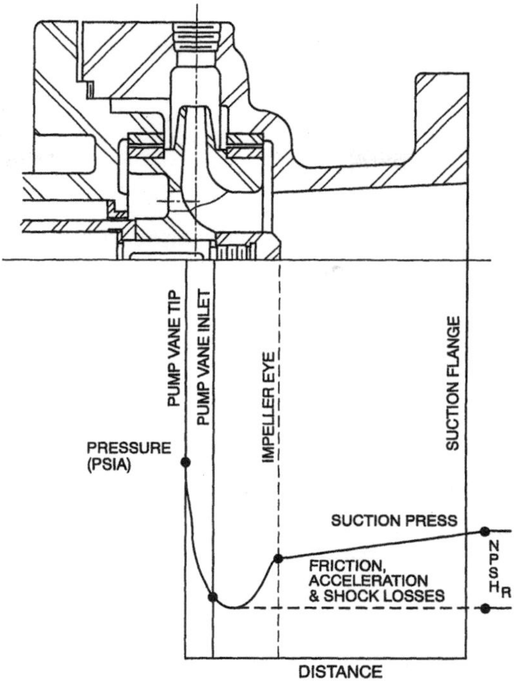

Figure 8.17 Flange to vane entrance losses

However, it must be remembered that $NPSH_A \geq NPSH_R$ is only one of the requirements that must be met to prevent vaporization. The following causes of vaporization must also be prevented:

- Low velocity stall
- Low flow temperature rise

Low flow temperature rise and fluid vaporization can be determined by the relationship shown in Figure 8.11 and the pumped fluid characteristics. However, the determination of low velocity stall or recirculation requires the understanding of the concept of suction specific speed.

Suction specific speed

N_{ss}, known as suction specific speed is determined by the same equation used for specific speed N_s but substitutes $NPSH_R$ for H (pump head). As the name implies, N_{ss} considers the inlet of the impeller and is related to the impeller inlet velocity. The relationship for N_{ss} is:

$$N_{SS} = \frac{N\sqrt{Q}}{(NPSH_R)^{3/4}}$$

Where: N = speed

Q = flow-GPM

$NPSH_R$ = Net Positive Suction Head Required

As previously explained, $NPSH_R$ is related to the pressure drop from the inlet flange to the impeller. The higher the $NPSH_R$, the greater the pressure drop. The lower the $NPSH_R$, the less the pressure drop. From the equation above, we can show the relationships between $NPSH_R$, N_{SS}, inlet velocity, inlet pressure drop and the probability of flow separation in Figure 8.18.

NSS related to flow separation probability

N_{SS}	$NPSH_R$	Inlet velocity	Inlet passageΔP	Probability of flow separation
14,000 (High)	Low	Low	Low	High probability
8,000 (Low)	High	High	High	Low probability

Figure 8.18 N_{SS} related to flow separation probability

Based on the information presented in Figure 8.18, it can be seen that flow separation will occur for high specific speeds resulting from low inlet velocity. The critical question the pump user needs answered is 'At what flow does the disturbance and resulting cavitation occur'? This is not an easy answer because the unstable flow range is a function of the impeller inlet design as well as the inlet velocity. A general answer to this question is shown in Figure 8.19.

Recirculation as a function of N_{SS}

The onset flow of recirculation increases with increasing suction specific speed

Figure 8.19 Recirculation as a function of N_{SS}

Another way of describing the statement in Figure 8.19 is 'The higher the value of N_{SS}, the sooner the pump will cavitate when operating at flows below the BEP'. Therefore, before an acceptable value of N_{SS} can be determined, the process system and pumped liquid characteristics must be defined.

Defining the process system

Reviewing the proposed process system prior to the purchase of a pump, as previously discussed, is strongly recommended. Figure 8.20 presents a typical process system with various control alternatives (flow, level, pressure).

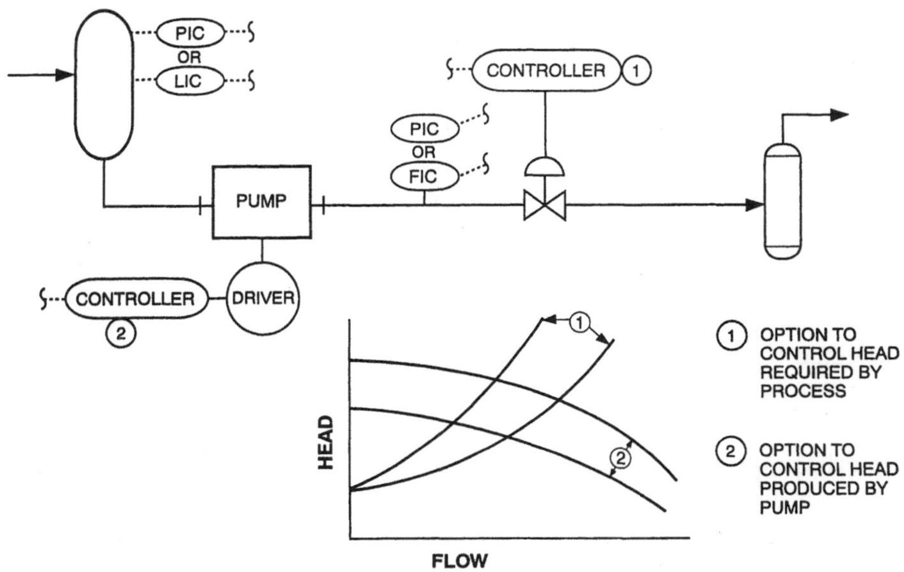

Figure 8.20 Centrifugal pump control options

The approach that should be followed when purchasing a pump is to define the required operating range of the pump based on the process system design and process requirements.

Once the operating range is defined, hydraulic calculations will determine the required flows, heads and $NPSH_{AVAILABLE}$. Care should be taken to define liquid composition and temperature as accurately as possible since these items will determine the vapor pressure which defines the $NPSH_{AVAILABLE}$. Steps in defining the process system are summarized in Figure 8.21.

Preventing liquid disturbances by accurately defining process requirements

- Define operating range of application
- Accurately define liquid characteristics
  - Vapor pressure
  - Pumping temperature
  - Viscosity
- Perform hydraulic calculations for all required flow rates to determine:
  - Head required
  - $NPSH_{AVAILABLE}$

Figure 8.21 Preventing liquid disturbances by accurately defining process requirements

Once the process requirements are accurately defined, a pump can be selected that will meet these requirements without the risk of hydraulic disturbances.

Selecting a pump for hydraulic disturbance free service

Having discussed the concepts used to prevent liquid disturbances and the requirements for accurately defining the process system requirements, this information can be used to select a pump free of hydraulic disturbances.

Refer to Figure 8.23 which shows a typical pump performance curve.

Based on previous discussions, there are three areas of concern to assure trouble-free operation (See Figure 8.22).

1 NPSH margin at maximum operating flow

2 Approximate recirculation margin at minimum operating flow

3 NPSH margin at minimum operating flow

Figure 8.22 Hydraulic disturbances – areas of concern

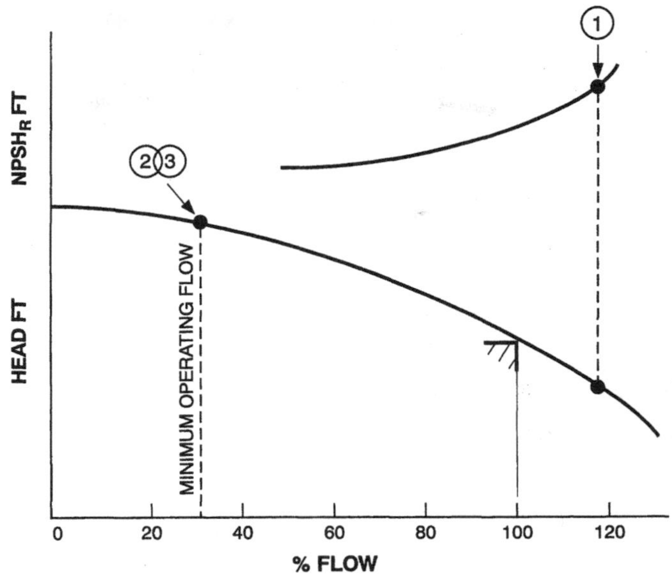

Figure 8.23 Hydraulic disturbance – areas of concern

The practical approach is to select a type of pump that will enable operation under all conditions in Figure 8.22 if possible. Figure 8.24 presents guidelines for selecting a pump free of hydraulic disturbances.

Guidelines for selecting pumps free of hydraulic disturbances

Step	Action
1.	Confirm $NPSH_A \geq NPSH_R$ at maximum operating flow. If margin less than two (2) feet, require witnessed $NPSH_R$ test. If $NPSH_R > NPSH_A$ ■ Increase $NPSH_A$ by: ● Increasing suction drum level ● Decreasing pumping temperature ● Decreasing suction line losses ■ Reselect pump (if possible) ■ Select canned pump
2.	For the pump selected calculate N_{ss} based on pump BEP conditions. Note: if double suction first stage impeller, use 1 /2 of bep flow
3.	If $N_{ss} > 8000$ contact pump vendor and require following data for actual pump fluid and conditions

- Predicted onset flow of cavitation caused by recirculation for actual fluid conditions
- Reference list of proposed impeller (field experience)

4. Compare cavitation flow to minimum operating flow. If this value is within 10% of minimum operating flow:
   - Reselect pump if possible
   - Install minimum flow bypass
   - Consider parallel pump operation
5. Calculate liquid temperature rise at minimum operating flow. If value is greater than 5% of pumping temperature:
   - Calculate $NPSH_A$ based on vapor pressure at calculated pumping temperature. If $NPSH_A < NPSH_R$
     - Install minimum flow bypass
     - Consider parallel pump operation

Figure 8.24 Guidelines for selecting pumps free of hydraulic disturbances

The guidelines presented in Figure 8.24 attempt to cover all situations, however, technical discussions with the pump vendor is encouraged whenever necessary.

Before proceeding, an important question regarding the typical pump performance curve needs to be asked. Why is the $NPSH_R$ curve not drawn to zero flow like the head curve? Based on the information presented in this course, you should be able to answer this question. Consider the following facts:

- The standard shop test fluid for all pumps is water
- The causes of vaporization at low flows

Hopefully your answer took the following 'form:

1. 'Liquid disturbances can occur at low flows if the vapor pressure of the pumped liquid exceeds the surrounding pressure of the liquid'
2. 'Flow separation and/or liquid temperature rise which can occur at low flows will either reduce the surrounding pressure on a liquid or increase its vapor pressure'
3. 'Since the actual liquid characteristics are not known when the standard pump curve (tested on water) is drawn, the vendor stops the $NPSH_R$ curve where flow separation and liquid temperature rise can cause liquid disturbances'

Therefore, trouble free operation to the left of this point is dependent on the pumped liquid and must be discussed with the pump vendor.

In conclusion, preventing liquid disturbances in the project design phase requires a thorough, accurate investigation of both the process and pump characteristics and some serious decisions on required action.

Justification of required action will be easier if the operating company looks beyond the project costs and examines what the actual cost (lost in production) will be if a pump experiences hydraulic disturbances. Most operating companies have documented Case Histories of problem pumps that will provide proven facts relating to the 'Total Cost' of operating problem pumps for the life of a project (Refer to Figure 8.25).

Determine the cost effectiveness of pump selection not only on the project (capital investment) costs, but on the cost to the operating company of unreliable pumps

Figure 8.25 Justifying the selection of troublefree pumps

Field operation

The preceding section described requirements to assure optimum pump reliability during the project design phase. This section will describe how to detect and correct hydraulic disturbances in the field.

Determining the potential for damage

Hydraulic disturbances are detected by monitoring the conditions noted in Figure 8.26.

Indicators of hydraulic disturbances

- Loud noise - continuous or varying
- Suction and discharge pressure pulsations
- High values of overall vibration
- Drop in produced head of >3%
- High values of vane passing frequency in overall spectrum
- Pressure pulsations in inlet and discharge piping
- Possible high bearing temperatures

Figure 8.26 Indicators of hydraulic disturbances

Once hydraulic disturbances are detected, the root cause of the disturbance must be defined. As previously discussed, vaporization of liquid must be present for cavitation to occur. Therefore, the requirement is to determine the root cause of vaporization. As previously discussed, there are three (3) primary root causes of vaporization:

- Internal inlet pressure losses
- Formation of low pressure cells at low flows
- Liquid temperature rise at low flows

Confirmation of process conditions and specific tests are required to determine the root cause of the problem.

Determining the cause of hydraulic disturbances

Confirmation of stated value of $NPSH_A$

Refer to the Pump Data Sheet and/or the Hydraulic Calculation Sheet for that pump service to determine the stated $NPSH_A$. Proceed to check the $NPSH_A$ at field operating conditions. Using the relationship:

$$NPSH_A = \frac{(P_{suction-PSIA} - P_{vapor\ pressure-PSIA}) \times 2{,}311}{S.G.\ at\ pumping\ temperature}$$

Substituting in the equation the actual values as follows:

- Pump suction pressure (down stream of suction streiner)
- Actual vapor pressure at measured pumping temperature
- Actual S.G. at measured pumping temperature

Compare calculated $NPSH_A$ to predicted $NPSH_A$. If actual $NPSH_A <$ predicted $NPSH_A$ modify operation if possible to attain predicted value. If this is not possible, the following alternatives exist:

- Operate pump at lower flow rate to reduce $NPSH_R$
- Modify pump to reduce $NPSH_R$ at operating flow rate

A summary of this discussion is presented in Figure 8.27.

Confirmation of $NPSH_A$ and recommended action

- Obtain predicted value of $NPSH_A$
- Calculate $NPSH_A$ using actual conditions
- If $NPSH_A$ actual < $NPSH_A$ predicted
  - Increase $NPSH_A$ if possible
  - Operate pump at lower flow rate (if cost effective)
  - Modify pump to reduce $NPSH_R$

Figure 8.27 Confirmation of $NPSH_A$ and recommended action

Internal pressure loss test

The test outlined in Figure 8.28 will confirm if the liquid disturbance is caused by pump inlet pressure losses resulting from high liquid velocity.

Test to confirm high velocity cavitation

- Close pump discharge control valve to reduce flow
- If pump noise significantly reduces and conditions noted in Figure 8.26 become stable, cause is confirmed

Figure 8.28 Test to confirm high velocity cavitation

If high velocity cavitation is confirmed and the stated value of $NPSH_A$ is confirmed, possible solutions are presented in Figure 8.29.

- Increase $NPSH_A$ until quiet operation is achieved
- Reduce pump throughout if cost effective
- Operate two pumps in parallel
- Increase impeller eye area
- Impeller material change
- Purchase new pump with acceptable $NPSH_R$

Figure 8.29 High velocity cavitation solutions

Modification of impeller eye area is not always possible and the pump vendor must be consulted to confirm it is possible and that satisfactory results have been achieved.

Before leaving this subject, mention of the inlet piping arrangement is required. Suggested inlet piping arrangements are presented in Figure 8.30.

Failure to conform with the guidelines presented in Figure 8.30 can lead to hydraulic disturbances caused by:

- Entrained vapor
- Additional internal ΔP caused by turbulence
- Liquid separation from impeller vanes

Inlet piping arrangements to avoid hydraulic disturbances caused by process piping

- Piping runs directly vertically or horizontally into pump without high pockets that can cause vapor formation
- Minimum straight suction pipe runs of:
  - Three (3) pipe diameters – single suction
  - Five (5) pipe diameters – double suction
- Double suction pumps should have pipe elbows perpendicular to the pump shaft
- The 'belly' of an eccentric reducer should be in the bottom location

Figure 8.30 Inlet piping arrangements to avoid hydraulic disturbances caused by process piping

Low flow hydraulic disturbance test

The test outlined in Figure 8.31 will determine if the liquid disturbance is caused by low flow circulation or temperature rise.

Test to confirm low flow hydraulic disturbances

- Open pump discharge control valve or bypass to increase flow
- If pump noise significantly reduces and conditions noted in Figure 8.26 become stable, cause of either low flow recirculation or temperature rise cavitation is confirmed
- Calculate liquid flow temperature rise to confirm if recirculation is the root cause

Figure 8.31 Test to confirm low flow hydraulic disturbances

Possible solutions are presented in Figure 8.32.

Solutions – low flow hydraulic disturbances

- Increase pump flow rate, if possible, until quiet operation is achieved
- Modify inlet volute to increase impeller inlet velocity (if sufficient $NPSH_A$ exists)
- Install impeller with reduced eye area (assuming-sufficient $NPSH_A$ exists) *
- Install minimum flow bypass to increase flow rate (to eliminate low flow temperature rise)

* Note: wear ring modifications are required

Figure 8.32 Solutions – low flow hydraulic disturbances

Internal volute and/or impeller eye modifications are not always possible. The pump vendor must be consulted to confirm these modifications are acceptable and satisfactory results have been achieved.

Justification of proposed action plan

Regardless of the cause of hydraulic disturbances, the problem cannot be resolved without management endorsement of a cost effective action plan. As previously mentioned, all action plans must be justified by cost savings. In the field, the largest revenue loss is usually lost product revenue resulting from unplanned critical equipment downtime.

Regardless of the proposed action, be sure to show lost product revenue against capital investment for problem solution.

9 Centrifugal pump testing

- Introduction
- Shop testing
- Field pump hydraulic and mechanical condition determination

Introduction

Once a centrifugal pump is selected, designed and manufactured for a specific duty, the shop test represents the final check available to the purchaser to confirm the pump will meet all hydraulic and mechanical requirements. In addition, the shop test represents the hydraulic and mechanical signatures or "baseline conditions" of all parameters that will be monitored for the field predictive and preventive maintenance programs.

This chapter will present details concerning shop testing of centrifugal pumps and methods for determining pump hydraulic and mechanical condition in the field.

Shop testing

Regardless of the type of pump selected, a certified pump performance test is recommended. It is the only checkpoint in the design and manufacturing cycle that will conclusively confirm proposed hydraulic and mechanical performance conditions have been satisfied. Refer to Figure 9.1.

Requirements for pump performance tests

To assure pump safety and reliability, all pumps should be performance tested

Figure 9.1 Requirements for pump performance tests

Many industry specifications do not require that every pump be performance tested. However, when the costs of reduced or lost product revenue are considered, the shop testing of all pumps to be used in the refining, petrochemical and gas processing industries can be justified.

While it is recommended that all pumps be tested, it is not recommended that they be witness tested unless they are in the categories noted in Figure 9.2.

Recommended basis for centrifugal pump witness tests

All pumps should receive certified performance tests. They should be witness tested if:

- They are of prototype design
- Critical service applications (boiler feed, re-injection, cooling water)
- An $NPSH_R$ test is required

Figure 9.2 Recommended basis for centrifugal pump witness tests

Typically, a pump shop test is called a performance test and includes determination of both hydraulic and mechanical characteristics. The practice is different than shop tests of compressor, steam and gas turbines which call for separate performance tests and mechanical tests to determine aerodynamic and mechanical characteristics. Pump performance tests determine that the hydraulic and mechanical parameters noted in Figure 9.3 are within specified limits.

Centrifugal pump test parameters

- Head
- Efficiency
- $NPSH_R$*
- Flow rate
- Overall vibration
- Bearing temperature rise
- Leak free seal operation**

*when specified by purchaser
**assumes mechanical seal installed

Figure 9.3 Centrifugal pump test parameters

Test set-up and requirements

Every pump test must be based on a specified test procedure which should contain the items noted in Figure 9.4.

Recommended test procedure content

- Test set-up schematic (arrangement, instruments and locations)
- Calibration requirements
- Measured parameters
- Acceptance limits (each measured parameter)
- Method of data reduction

Figure 9.4 Recommended test procedure content

Set-ups for closed and open loop tests are shown in Figures 9.5 and 9.6. The type of loop used depends on test requirements and vendor facilities. A closed loop must be used when $NPSH_R$ tests are required for pumps that have large $NPSH_R$ values or if water is not used as the test fluid.

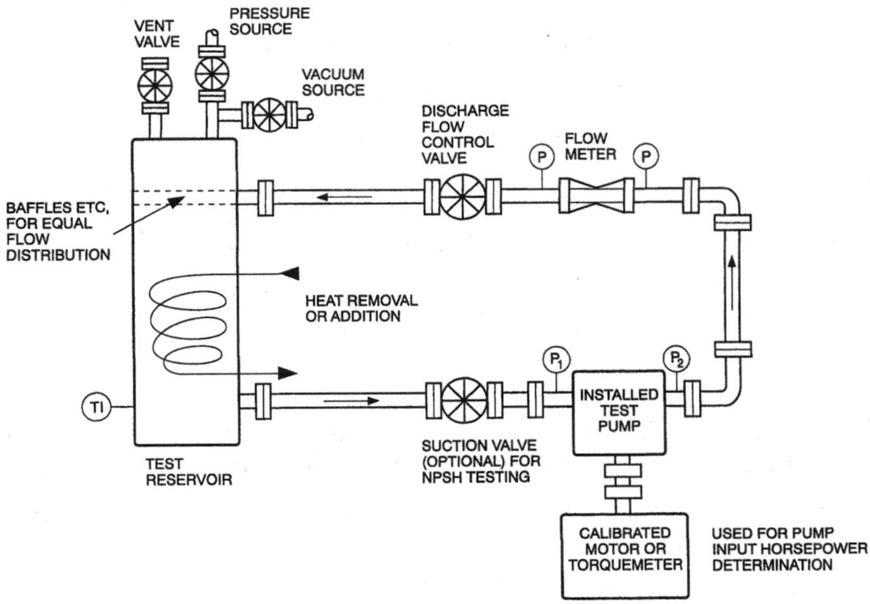

Figure 9.5 Closed loop pump test schematic

Clean water with a rust inhibitor is the universally accepted test liquid in either an open or closed loop. We have found that there is much confusion in the industry regarding the validity of a shop test curve based on water that will be used on a different liquid in the field. The principles learned thus far in this course should eliminate all confusion in the matter. Refer to Figure 9.7.

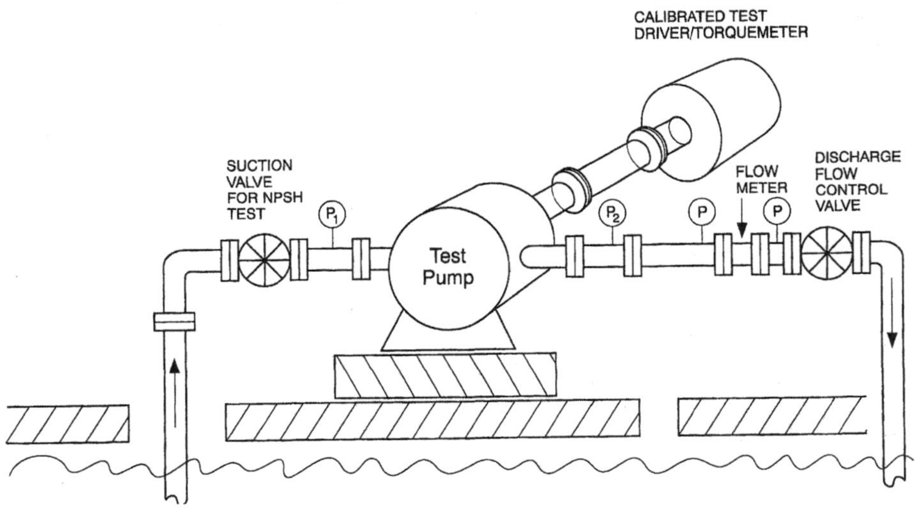

Figure 9.6 Open loop pump test schematic

Validity of head curves for any non-viscous liquid

A pump head (energy) vs flow curve at a specified speed is valid regardless of the type of non-viscous liquid pumped because:

- Head is produced only by:
  - Vane tip speed
  - Tangential velocity (function of relative velocity)
- All liquids are incompressible

Figure 9.7 Validity of head curves for any non-viscous liquid

The objective of all pump head vs flow tests is to verify that the tested pump produces the required head at a given flow rate and pump speed.

Pump efficiency is also independent of the pumped non-viscous liquid since it is directly related to velocity.

Horsepower, which is a function of head, efficiency and mass flow is dependent on the pumped liquid and varies directly as the ratio of:

$$\frac{\text{S.G. pumped liquid}}{\text{S.G. of test liquid (1.0)}}$$

All pump curves tested on H_2O must be corrected by this ratio to determine actual drive horsepower required.

$NPSH_R$ is determined by change in pump produced head and is a function of the pumped liquid. Therefore, the $NPSH_R$ will change slightly as a function of the pumped fluid. Consult the pump vendor if the margin between $NPSH_{AVAILABLE}$ and $NPSH_{REQUIRED}$ is less than 3ft (1 meter)

A summary of this discussion is presented in Figure 9.8.

H_2O test curve field validity

A pump test curve based on H_2O

Is completely valid for any non-viscous liquid

- Head curve
- Efficiency curve

Must be corrected for pumped liquid

- Horse power curve (corrected by pumped S.G.)
- $NPSH_{REQUIRED}$ (slight correction – consult pump vendor)

Figure 9.8 H_2O test curve field validity

Instrument calibration

All test instruments must be calibrated. Calibration corrections if any, must be included in final calculations. Industry test codes require that instruments, including flow meters must be periodically calibrated.

Acceptance limits on measured parameters

Acceptance limits vary slightly depending on the type of centrifugal pump. In Figure 9.9 we have listed suggested acceptance limits for a single stage pump and a multi-stage pump based on the measured parameters noted in Figure 9.3.

Suggested centrifugal pump test acceptance limits

	Allowable limits for rated flow	
Parameter	Single stage	Multi-stage
Head	-2%	-2%
	+4%	+2%
Efficiency	varies - not to exceed head and H.P. requirements	
Horsepower	+4%	+4%
$NPSH_R$	+0%	+0%
Overall vibrations	0.2 in/.sec.	0.2 in/sec.
Bearing bracket (anti-friction bearings)	180°F	N.A.
Bearing temperature rise (hydrodynamic drain ΔT bearing)	N.A.	60°F ΔT

Figure 9.9 Suggested centrifugal pump test acceptance limits

Hydraulic and mechanical tests

Once a test set-up is selected and all instruments to be used are calibrated, the pump performance test is conducted. A basic performance test outline for data acquisition and data reduction is presented in Figures 9.10 and 9.11.

Performance test outline – data acquisition

For each data point (usually 6–8 points including shut-off) obtain the following:

- $P_{INLET\ TOTAL}$ (P_{STATIC} AND $P_{VELOCITY}$)*
- $P_{DISCHARGE\ TOTAL}$ (P_{STATIC} AND $P_{VELOCITY}$)*
- Flow (GPM)
- Speed
- Test liquid S.G.
- Driver power
- Liquid temperature
- Vibration (each bearing)
- Bearing temperature (each bearing)

$$*P_{VELOCITY\ (FT)} = \frac{V^2}{2g}$$

where: V = velocity of liquid in pipe-ft/sec
g = gravitational constant 32 ft/sec^2

Figure 9.10 Performance test outline – data acquisition

Performance test outline – data reduction

For each data point reduce data as follows:

- Flow $$\text{GPM} = (\text{flow meter } \Delta P) \times \text{correction factor} \times \left(\frac{N_{RATED}}{N_{TEST}}\right)^{*}$$
- Head $$\text{Head} = \left[\frac{2.311 \times (P_{2TOTAL} - P_{1\ TOTAL})}{\text{S.G.}}\right] \times \left(\frac{(N_{RATED})}{N_{TEST}}\right)^{2*}$$
- Efficiency $$\text{Eff'y} = \frac{\text{Hd} \times \text{GPM} \times \text{S.G.}}{3960 \times \text{BHP}_{MEASURED}}$$
- BHP $$\text{BHP} = \frac{\text{Head}_{RATED\ SPEED} \times \text{GMP}_{RATED\ SPEED} \times \text{S.G.}_{RATED}}{3960 \times \text{eff'y}}$$

*when test speed ≠ rated speed

Figure 9.11 Performance test outline – data reduction

After all data is reduced, head, efficiency and BHP vs flow are plotted on a certified test curve.

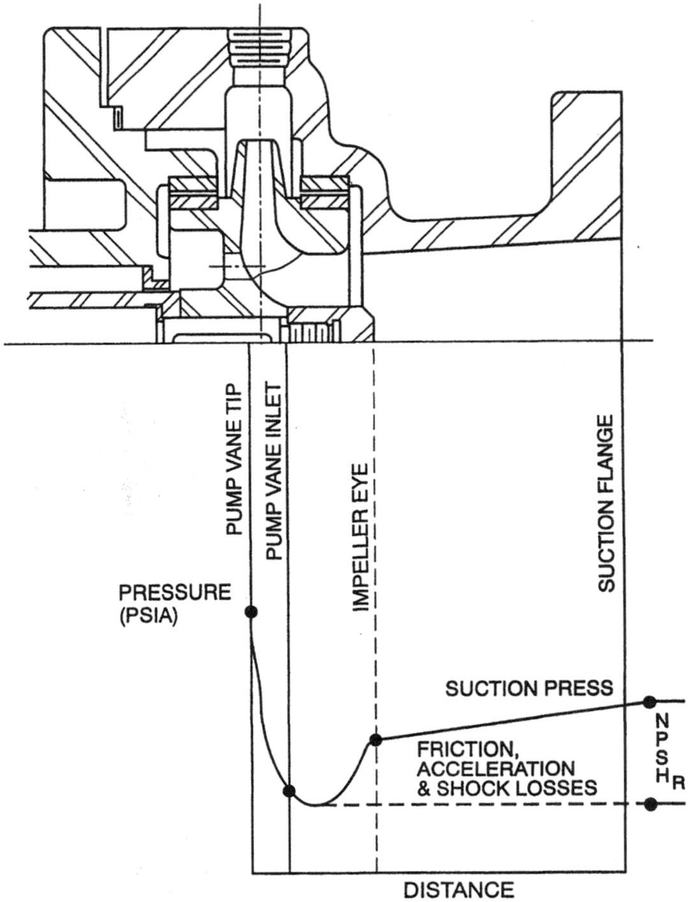

Figure 9.12 Flange to vane entrance losses

$NPSH_{REQUIRED}$ test

When requested or when a new impeller design is tested, a Net Positive Suction Head Test is performed. The test set-up is basically the same as for a performance test with the exception that a source of suction pressure suppression must be provided. The available sources are:

- Variable suction vessel pressure
- Suction valve

The objective of a NPSH test is to determine the point where cavitation begins. As previously discussed, cavitation first requires that a vapor be present. As shown in Figure 9.12, $NPSH_R$ is a function of the pressure drop in the suction passage of a pump.

If $NPSH_A$ is exactly equal to $NPSH_R$, the pressure drop in the suction passage will be exactly equal to the $NPSH_R$ expressed in feet and the

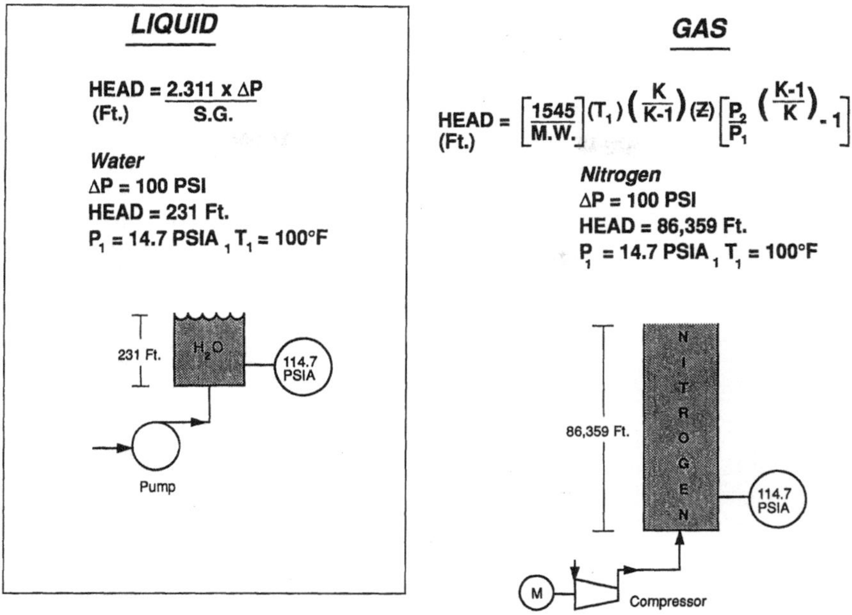

Figure 9.13 Fluid head. Left: liquid. Right: gas

pump will cavitate. As shown in Figure 9.13 when a pump cavitates, the head produced will decrease since the pumped fluid is now compressible and the head produced by the impeller is not sufficient to meet the head required at that flow rate. The problem in NPSH testing is; how much head drop defines cavitation? Strictly speaking, any head drop defines cavitation. However, considering normal hydraulic fluctuation encountered in any test loop, the pump industry has agreed on a 3% head drop while maintaining a constant flow as the definition of cavitation. However, it must be noted that in certain water applications – cooling water and boiler feed water, a 1% head drop has been used to define cavitation.

This practice is not recommended due to the difficultly in obtaining accurate test measurements. A recommended alternative for these applications is to maintain the standard 3% head drop and increase the NPSH margin to account for the application. This is precisely why we have recommended a 10 ft. margin for H_2O applications. Figure 9.14 outlines the NPSH test procedure.

$NPSH_R$ Test Procedure

Step 1. test pump to establish head vs flow curve

Step 2. for each flow point, reduce suction pressure keeping flow constant until head drops 3% from value in step 1.

Step 3. Calculate $NPSH_{REQUIRED}$ for 3% head drop at each flow point from $NPSH_R = \frac{P_{SUCTION\ TOTAL} \times 2.311}{S.G.}$

Figure 9.14 $NPSH_R$ test procedure

Figure 9.15 is a typical plot of a $NPSH_R$ test for different flow points as a function of $NPSH_R$ and pump head. For each flow, pump suction pressure is gradually reduced, thus reducing the $NPSH_R$. When the pump produced head decreases 3%, cavitation is present and the $NPSH_R$ at this point is plotted against flow which was held constant.

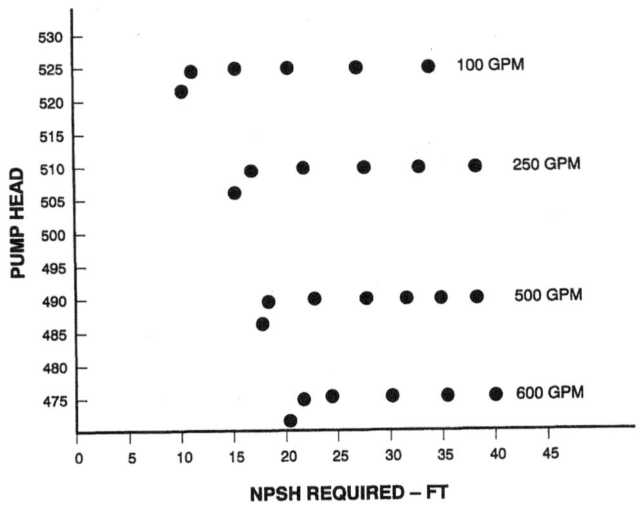

Figure 9.15 NPSH test plots

Field pump hydraulic and mechanical condition determination

Any effective field predictive and preventive maintenance program must be based on both hydraulic and mechanical condition monitoring. As previously discussed in this book, the root cause of many pump mechanical problems lies in hydraulic performance changes within the pump. For many reasons, hydraulic condition monitoring does not receive the same degree of attention as mechanical condition monitoring (vibration, bearing temperature, oil analysis, seal checks). Our experience has shown that operational requirements, insufficient instrumentation (particularly flow meters) and lack of knowledge regarding hydraulic performance are the major reasons for this discrepancy.

In this chapter, we will present practical methods to obtain centrifugal pump hydraulic performance parameters in the field to enable site personnel to determine the need for maintenance.

Centrifugal pumps unlike positive displacement pumps can experience flow rate reduction for reasons other than mechanical deterioration. As we have learned, positive displacement pumps for a given speed are constant flow pumps. If the flow rate decreases, a positive displacement pump requires maintenance.

However, centrifugal pump flow rate can change for either reason shown in Figure 9.16.

Reasons for centrifugal pump flow change

- Process head required increases
- Pump head produced decreases

Figure 9.16 Reasons for centrifugal pump flow change

In order to determine if pump head produced is decreasing, the pump suction and discharge pressures must be measured at a specific pump speed and flow rate. The problems in the field are that a calibrated flow meter is rarely if ever available, pressure gauges are not usually calibrated and the location of pressure gauges are not correct. Refer to Figure 9.17

Field performance condition monitoring requirements

The field is not a laboratory!

Field performance condition monitoring requires:

- Determination of pump flow
- Calibration of pressure gauges
- Proper location of pressure gauges

Figure 9.17 Field performance condition monitoring requirements

As noted in Figure 9.17, pressure gauges must be calibrated and properly located. Location of pressure gauges should be as shown in Figure 9.18.

Field test pressure gauge location requirements

- Pressure gauges must be located approximately one (1) pipe diameter from the pump flanges
- There must not be any restriction (orifice, strainer, etc.) between the pump flange and pressure gauge

Figure 9.18 Field test pressure gauge location requirements

The effectiveness of any predictive and preventive maintenance program depends on the proper location of instrumentation.

Methods for field determination of performance are presented below.

Pump performance determination procedure

As stated in Figure 9.17 any field performance procedure must determine flow rate. In addition, the test procedure cannot significantly impact plant operation. As stated, the field is not a laboratory. A six (6) point performance test is out of the question. The field test objective must be therefore to condition monitor and tend one (1) flow point.

How can this objective be achieved? Figure 9.19 presents field performance guidelines for centrifugal pumps.

Field performance guidelines – centrifugal pumps

1. Measure pump suction and discharge pressure using the pump to recycle back to the suction drum at either of the following operating points:
   A. Shut off flow – cold water pumps and low vapor pressure applications only
   B. Minimum flow – all other applications and whenever a minimum flow bypass line is available
2. Calculate pump head based on actual liquid specific gravity
3. Plot calculated head at either point A or B on shop test curve
4. If calculated head falls greater than 5% from curve, maintenance is required

Figure 9.19 Field performance guidelines – centrifugal pumps

The procedure outlined in Figure 9.19 allows performance determination of any installed pump without adversely impacting plant operating philosophy. It also enables performance determination without an installed flow meter since either shut-off (0 flow) condition or the minimum flow bypass line will assure performance is obtained at the same flow point.

Performance condition monitoring guidelines for an operating pump are presented in Figure 9.20.

Performance condition monitoring guidelines

- Establish baseline value of pump head and flow immediately after initial plant start-up
- Confirm acceptable trend of pump head at the original flow (< – 5%) prior to turnaround or if hydraulic disturbance is present
- Establish baseline value of pump head immediately after pump maintenance. Using the same guidelines as for initial plant start-up

Figure 9.20 Performance condition monitoring guidelines

In most cases, a flowmeter that directly measures flow from the specific pump in question will not be available. For these cases, Figure 9.21 presents alternatives to determine the flow rate of the pump.

Pump flow rate determination alternatives

- Obtain pump flow from control valve pressure drop, valve position and valve coefficient (obtained from valve engineering data for the specific valve). Note: control valve must handle all flow from the pump
- Use ultrasonic flow meter to measure pump flow
- Measure motor amps and calculate pump required power from the equation

$$\text{Pump power (bhp)} = \frac{(V)(A)(P.F.)(1.735)(\text{Motor efficiency})}{746}$$

Where: V = Volts to motor terminals
A = Amps to motor terminals
P.F. = Power factor (decimal)
1.735 = $\sqrt{3}$ = constant for 3 phase motor
Motor efficiency is expressed as a decimal

Plot calculated pump power on the pump shop test curve (note to correct curve for specific gravity (2)), to determine approximate flow

- Measure pipe inlet and outlet temperature and calculate pump efficiency from the following equation:

$$\Delta T = \frac{\text{Pump head}}{778 \times Cp} \times \left[\frac{1}{\text{Pump efficiency}} - 1\right]$$

Where: Pump head is calculated from data in $\frac{FT\text{-}LB_F}{LB_M}$

Cp is specific heat of the fluid in $\frac{BTU}{°F\text{-}LBmass}$

778 is conversion factor $\frac{FT\text{-}LB_F}{BTU}$

Pump efficiency is expressed as a decimal

Plot calculated pump efficiency on the pump shop test curve to determine approximate flow

Note: 1. Pump flow determination by amp and pipe temperatures assumes that pump is in original (new) condition

2. Correct horsepower curve for specific gravity by ratio of $\frac{\text{Actual S.G.}}{\text{S.G noted on curve}}$

Figure 9.21 Pump flow rate determination alternatives

10 Pump mechanical design

- Introduction
- Basic elements of centrifugal pumps
- Volutes
- Wear rings
- Impellers
- Bearings
- Anti-friction bearings
- Hydrodynamic bearings
- Balancing devices
- Shaft and key stress

Introduction

Regardless of the degree of pump performance optimization, the availability of any pump depends on the quality of its mechanical design and manufacture. Understanding the function of each mechanical component is essential in properly specifying, maintaining and operating any pump.

Basic elements of centrifugal pumps

Each type of centrifugal pump is made up of two (2) basic elements, stationary and rotating (Refer to Figure 10.1).

Basic elements of centrifugal pumps

I. Casing

- Provides nozzles to connect suction and discharge piping
- Directs flow into and out of impeller and converts kinetic energy into pressure energy
- Provides support to the bearing housing

II. Bearings and stuffing box

- Provides support and enclosure for the rotating element

Figure 10.1 Basic elements of centrifugal pumps

In addition, any pump is comprised of hydraulic and mechanical components. The first prt of this chapter will cover the hydraulic components (casing, volute, diffusers, wear rings and impellers. The remainder of this chapter will cover the mechanical components (bearings and seals). The major components which make up these elements, and their functions, will be described.

Volutes

The volute is that portion of the pump casing where the liquid is collected and discharged by centrifugal force when it leaves the impeller (Refer to Figure 10.2).

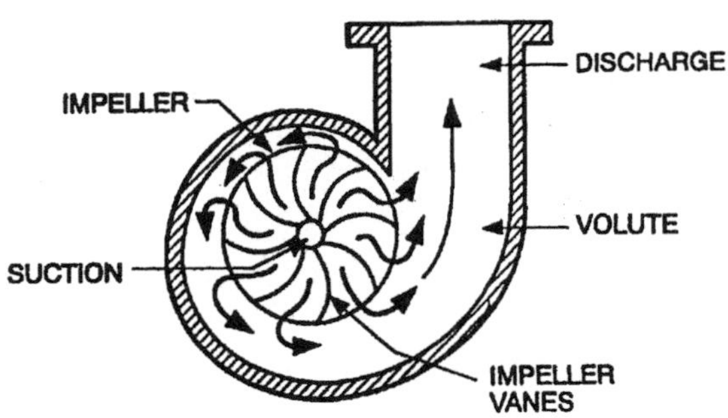

Figure 10.2 Single volute casing form

As the liquid leaves the rotating impeller the volute continually accumulates more liquid as it progresses around the casing. Because we want to keep the liquid velocity reasonably constant as the volume increases, the volute area between the impeller tip and the casing wall must be steadily increased since:

$$Q = AV$$

Where Q = Flow (gpm)

A = Area

V = Velocity

Figure 10.2 shows this relationship.

The form of volute casing shown in Figure 10.2 results in uneven pressure distribution around the periphery of the impeller. In a single stage overhung impeller type pump with cantilever shaft designed for high heads, the unbalanced pressure may result in increased shaft deflection and bearing loadings at off-design conditions.

The double volute casing design shown in Figure 10.3 equalizes the pressure around the impeller and reduces the unbalanced loading on the bearings. This is accomplished with two (2) similar flow channels which have outlets 180 degrees apart.

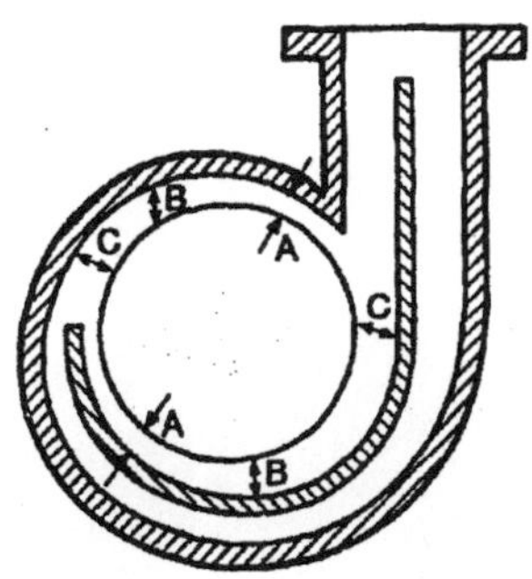

Figure 10.3 Dual volute casing form

Figure 10.4 shows the relationship between the radial reaction force acting on an impeller vs capacity for single and double volute pumps. The radial reaction force is directly proportional to the following parameters:

- Impeller produced head
- Impeller diameter
- Impeller discharge width (b_2)
- Specific gravity

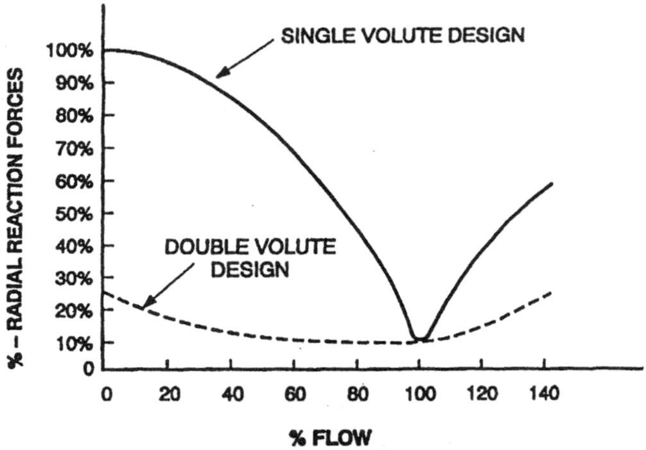

Figure 10.4 Volute radial reaction forces versus capacity

To balance the pressure around the periphery of the impellers of a multistage volute casing type pump, staggered volutes or double volute designs are available, when size permits.

Some multistage pumps, particularly the radial split barrel type casing design, utilize vaned diffusers to convert kinetic energy to pressure energy. The vanes are designed such that the liquid flow area gradually increases to effect a gradual decrease in velocity from inlet to outlet (Refer to Figure 10.5).

Figure 10.5 Vane type diffuser

Wear rings

To minimize the amount of liquid leakage from the high pressure side to the suction side of a centrifugal pump, wear rings are usually fitted to the casing and also to the suction side of the impeller eye outside diameter. To further minimize the leakage, wear rings can be fitted to the casing cover inside diameter and to the back side of the impeller (Refer to Figure 10.6).

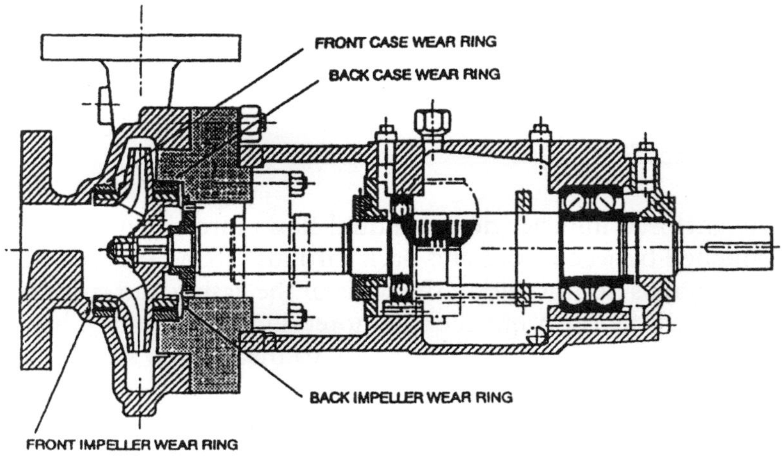

Figure 10.6 Wear ring locations

The clearance provided is small and the material hardness of the case and impeller wear rings is slightly different to avoid potential galling of the parts. The wear rings are lubricated by the pumped liquid. As wear takes place over time, clearance will increase and more flow will pass back to suction, resulting in some degradation of efficiency. API Standard 610 recommends clearances for these parts which are slightly more liberal than would otherwise be provided by most manufacturers. When a pump is furnished with API clearances, a slight decrease in efficiency is accounted for in pump performance.

Impellers

The impeller can be considered the heart of the centrifugal pump. It is the only component which produces head (energy). Therefore, they should be well designed to handle the liquid pumped with minimum loss. There are two (2) basic impeller designs available; the enclosed type, and the open type (Refer to Figure 10.7).

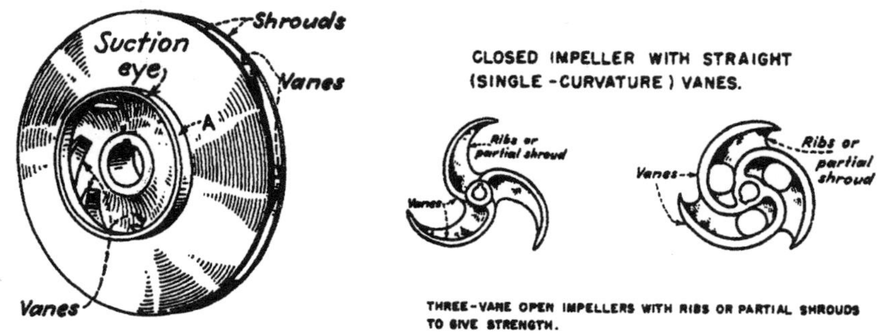

CLOSED IMPELLER WITH STRAIGHT (SINGLE -CURVATURE) VANES.

Figure 10.7 Impeller designs

In the enclosed impeller design, liquid flows into the impeller where it is contained between front and back shrouds affixed on each side of the vanes. There is no flow along the walls of the casing and suction head as there is with the open impeller. As noted in the previous Chapter, flow circulation from the high pressure side to the suction side is minimized by the use of close clearance wear rings.

The open impeller design may have vanes cast integral with the hub, usually without side walls. This type of design requires that the impeller be positioned with close clearance between it and the casing wall. This close clearance operation minimizes recirculation of liquid from the area under discharge pressure to the area under suction pressure. The efficiency of the open impeller is lower than that of the enclosed type. If the clearance between the impeller vane and the casing wall increases, some adjustment, to close the clearance, is possible. The open impeller does not find widespread use in hydrocarbon service.

The impeller may be further classified as single suction or double suction type. The term single and double suction defines the number of inlets to the impeller. A single suction type has one (1) inlet and a double suction type has two (2) inlets (Refer to Figure 10.8).

The larger the suction area in a double suction impeller allows the pump to operate with less positive suction head for a given capacity as compared to a single suction impeller because the flow area to the impeller vanes is increased. This results in a reduced pressure drop from the pump suction flange to the vanes in the eye of the impeller and thus reduces the NPSH required.

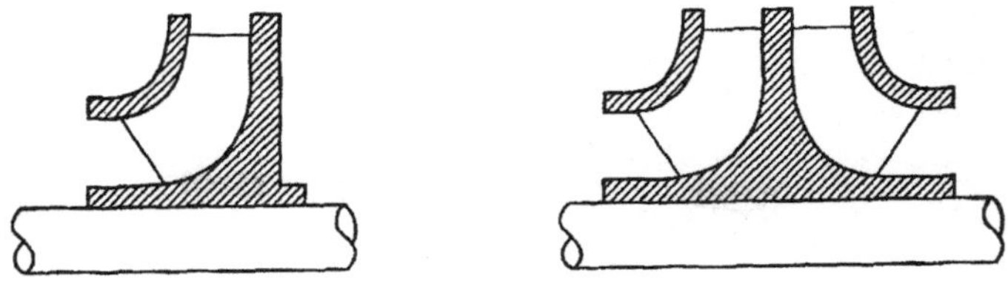

Figure 10.8 Impeller suction types

Specific speed

Specific speed is a non-dimensional value that is a function of pump speed, flow and head.

$$Ns = \frac{N\sqrt{Q}}{Hd^{(3/4)}}$$

Where: N_S = specific speed

N = pump speed RPM

Q = pump flow GPM

Hd = pump produced head $-\frac{FT\text{-}lb_f}{lb_m}$

Specific speed is use extensively in both pump and compressor design to optimize stage efficiency for a given value of flow and head required. In pump design, specific speed is used to optimize the following design parameters:

- Impeller discharge flow velocity
- Impeller tip speed
- Impeller inlet and discharge blade angles
- Discharge throat velocity

Figure 10.9 presents a plot of pump specific speed vs efficiency for various flow rates. This is a typical chart and will vary slightly from pump vendor to pump vendor based on the specifics of pump design. It can be used for estimating purposes in determining pump required horsepower.

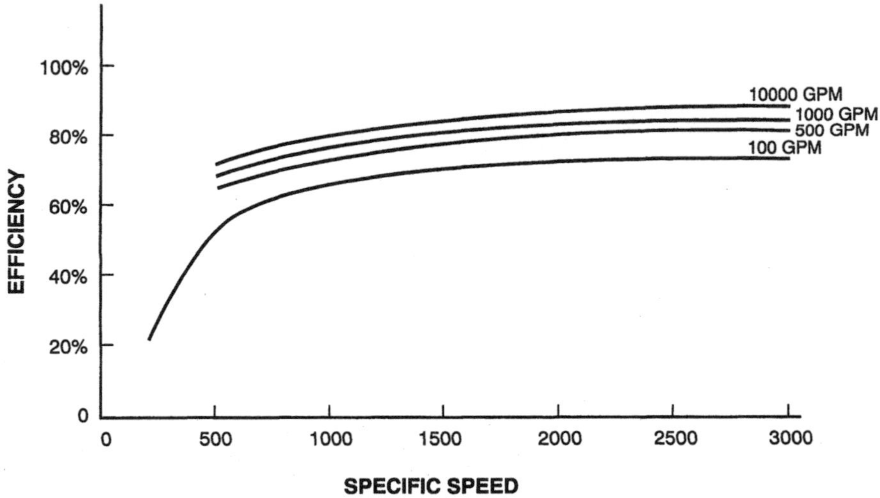

Figure 10.9 Efficiency as a function of pump specific speed and flow

Bearings

Functions of bearings

Bearings provide support for the rotating element. They are required to carry radial and axial loads. The basic relationship for any bearing design is:

$$P = \frac{F}{A}$$

Where: P = Pressure on the supporting oil film (hydrodynamic) or element (anti-friction) in P.S.I.

F = The total of all static and dynamic forces acting on the bearing in LB force

A = The bearing area in IN^2

In many cases, a pump bearing may perform for years and suddenly fail. This abrupt change in performance characteristics is usually due to a change in the forces acting on the bearing. Factors that can increase bearing load are shown in Figure 10.10.

Sources of forces

- Increased process pipe forces and moments
- Foundation forces ('soft' foot, differential settlement)
- Fouling or plugging of impeller
- Misalignment
- Unbalance
- Rubs
- Improper assembly clearances
- Thermal expansion of components (loss of cooling medium, excessive operating temperature)
- Radial forces (single volute – off design operation)
- Poor piping layouts (causing unequal flow distribution to the pump)

Figure 10.10 Sources of forces

In this section we will discuss the functions and various types of bearings used in pumps.

There are three (3) basic types of bearings used in centrifugal pumps (Refer to Figure 10.11).

Basic types of bearings

- Anti-friction
- Hydrodynamic ring oil lubricated
- Hydrodynamic pressure lubricated

Figure 10.11 Basic types of bearings

Application guidelines

Centrifugal pumps in the process industry are normally fitted with bearings which are appropriate for the application and pump design (Refer to Figure 10.12).

The L-10 rating life for anti-friction bearings is defined as the number

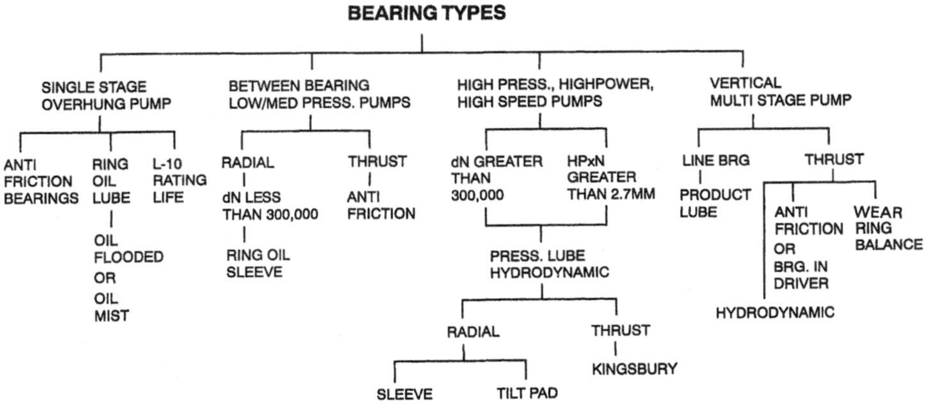

Figure 10.12 Bearing application guidelines

of hours at rated bearing load that 90% of a group of identical bearings will complete or exceed (25,000 hours of operation) before the evidence of failure. Failure evidence is generally defined as a 100% increase in measured vibration.

API Standard 610 recommends applying pressurized hydrodynamic bearings when the product of pump rated horsepower and rated speed in revolutions per minute exceeds 2.7 million. Pressure lubrication systems may be integral or separate, but should include as a minimum an oil pump, reservoir, filter, cooler, controls and instrumentation.

The dN number is the product of bearing size (bore) in millimeters and the rated speed in revolutions per minute. It is used as a measure of generated frictional heat and to determine the type of bearing to be used and type of lubrication required. Figures 10.13 and 10.14 present the relationships for these two important factors.

L-10 Life
Used for all anti-friction bearings

$$L-10 = \frac{16700}{N} \bullet \left[\frac{C}{F}\right]^3$$

where: L-10 = hours of operation 90% of a group of identical bearings will complete or exceed

N = pump speed

C = the total force required to fail the bearing after 1,000,000 revolutions (failure defined as 100% increase in measured vibration)

F = the total of all actual forces acting on the bearing

Figure 10.13 L-10 Life

D-N Number

- Is a measure of the rotational speed of the anti-friction bearing elements
- D-N number = bearing bore (millimeters) X speed (RPM)
  Is used to determine bearing type and lubrication requirements

D-N range	bearing type	lubrication type
Below 100,000	anti-friction	sealed
100,000 - 300,000	anti-friction	regreasable
Below 300,000	anti-friction	oil lube (unpressurized)
Above 300,000	sleeve, multi-lobe or tilt pad	oil lube (pressurized)

Figure 10.14 D-N Number

Anti-friction bearings

Ball Bearings

The anti-friction bearing most commonly used in centrifugal pumps for carrying radial and thrust loads is the ball bearing. The design of each type of bearing has it's individual advantages for specific load carrying requirements (Refer to Figure 10.15).

TYPE	LOAD CAPACITY	DESCRIPTION
• SINGLE-ROW DEEP GROOVE	EQUAL THRUST CAPACITY IN EACH DIRECTION. MODERATE TO HEAVY RADIAL LOADS	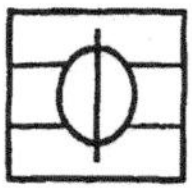
• SINGLE-ROW ANGULAR CONTACT	HIGHER RADIAL LOAD THAN DEEP GROOVE. HEAVY THRUST LOAD IN ONE DIRECTION	
• DUPLEX, SINGLE-ROW BACK TO BACK (DB)	HEAVY RADIAL LOADS, COMBINED RADIAL/THRUST LOADS, REVERSE THRUST	

Figure 10.15 Typical ball bearing load capacity guidelines

Roller bearings

Anti-friction roller bearings are capable of accepting pure radial loads, pure thrust loads or various combinations. They have been applied as crankshaft bearings in power ends of some power pump designs. Refer to Figure 10.16. Specific uses of this type of bearing resides with the manufacture and is not usually a selective option for the user.

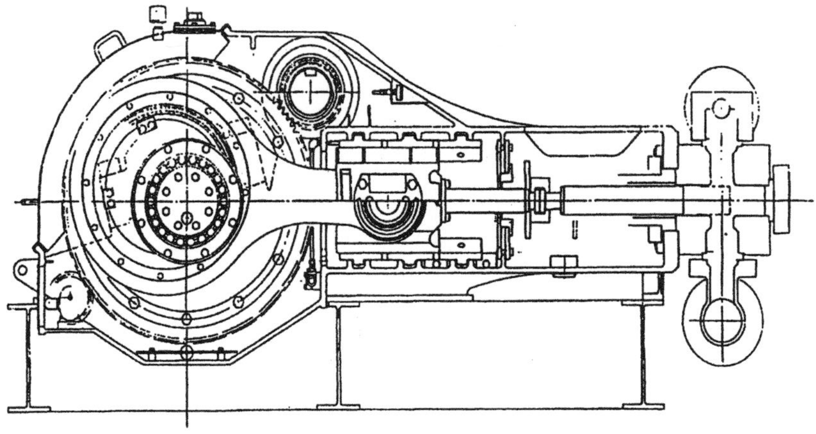

Figure 10.16 (Courtesy of Union Pump Co)

Lubrication

The purpose of anti-friction bearing lubrication is to increase bearing life, keep the balls or rollers separated, dissipate the heat generated in and conducted into the bearing and to prevent corrosion. There are various methods of oil lubrication for ball and roller bearings with the two most common methods being the ring oil and flooded method. Refer to Figures 10.17 and 10.18.

A word of caution – both types of lubrication usually incorporate constant level oilers to maintain oil level at a specified height in the bearing housing. The oil level in the constant level oiler is not the level of the oil in the bearing housing. The constant level oiler is a reservoir that provides oil to maintain a constant level in the bearing housing.

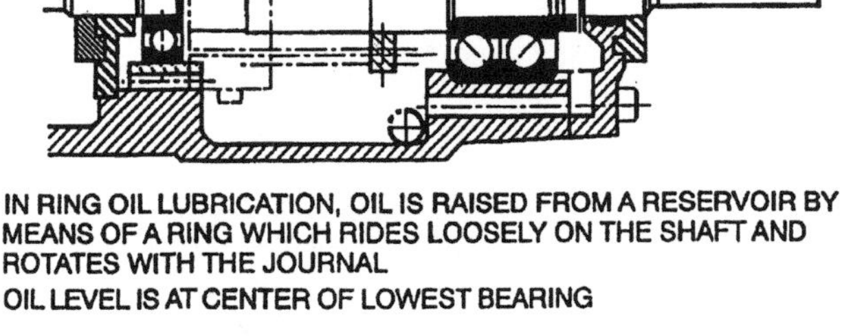

- IN RING OIL LUBRICATION, OIL IS RAISED FROM A RESERVOIR BY MEANS OF A RING WHICH RIDES LOOSELY ON THE SHAFT AND ROTATES WITH THE JOURNAL
- OIL LEVEL IS AT CENTER OF LOWEST BEARING

Figure 10.17 Ring oil lubrication

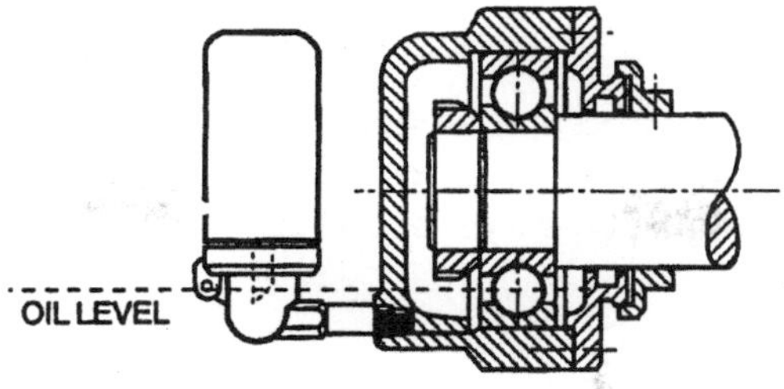

• IN FLOODED LUBRICATION, LUBRICATION IS ACHIEVED BY MAINTAINING AN OIL LEVEL IN THE RESERVOIR AT ABOUT THE CENTER OF THE LOWEST BEARING
• CONSTANT LEVEL OILER MAINTAINS LEVEL

Figure 10.18 Flood type lubrication

Proper oil level must be confirmed by visual inspection or by using a dip stick.

Oil mist lubrication is another acceptable method of lubricating anti-friction bearings. It is gaining acceptance since it provides a controlled environment in the bearing housing in both operating and non operating pumps. A mist generator console is often used to provide services to many pumps simultaneously. Use of oil mist lubrication should be investigated in environments where sand or salts exist. (Refer to Figure 10.19).

Mist oil system features

Features of oil mist system include:

- Mist generator provides mixture of air and atomized oil under pressure
- Reduced operating temperature resulting from air flow preventing excessive Oil accumulation
- Oil consumption is low. controlled with instrumentation
- Entrance of grit and contaminants is prevented since air is under pressure
- Reliable continuous supply of lubricant is available
- Ring oil can be installed as back-up to mist system (called "wet sump system")

Figure 10.19 Mist oil system features

Hydrodynamic bearings

The function of the hydrodynamic bearings is to continuously support the rotor with an oil film that is less than one (1) thousandth of an inch. There are various types of bearings available (Refer to Figure 10.20).

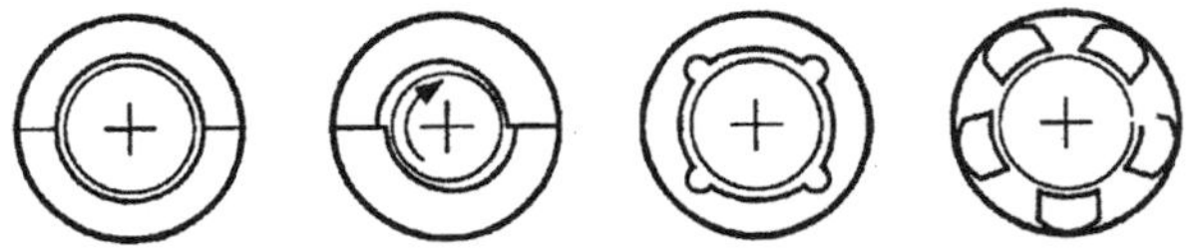

Figure 10.20 Journal bearing types

Sleeve bearings

Sleeve bearings are commonly used as radial bearings in centrifugal pumps. Lubrication is usually supplied by an external pressurized system, although some slow speed sleeve bearing applications (dN less than 400,000) can utilize ring oil lubrication.

Tilt Pad bearing

Tilt pad journal bearings are sometimes used for high horsepower, high speed centrifugal pumps where rotor stability may be of concern. They are always oil pressurized bearings.

Multilobe bearing

This type of bearing has found limited use in centrifugal pumps, but it can also be applied on high speed pumps where increased damping and stiffness is desired for lightly loaded bearings (most vendors will select tilt pad bearings).

Hydrodynamic bearing performance

When at rest, the journal settles down and rests at the bottom of the bearing (Refer to Figure 10.21A). As the journal begins to rotate, it rolls up the left side of the bearing, moving the point of contact to the left. There is then a thin film of oil between the contact surfaces, and fluid friction takes over for metal to metal contact (Refer to Figure 10.21B). The journal slides and begins to rotate, dragging more oil between the surfaces, forming a thicker film and raising the journal. As the speed of rotation increases, the oil drawn under the journal builds up pressure that forces the journal up and to the right in the direction

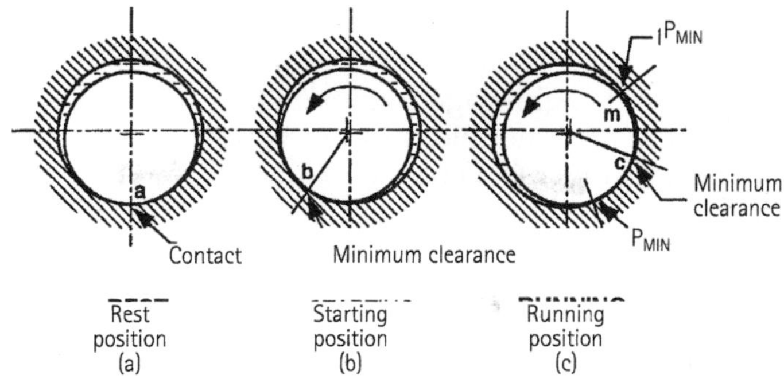

Figure 10.21 Hydrodynamic bearing performance

of rotation shown until a condition of equilibrium is reached resulting in a point of minimum clearance (Refer to Figure 10.21C).

As the oil film builds up under the journal, the center of the journal moves and the location of minimum clearance is away from the load line. The pressure distribution across the bearing varies and the maximum unit pressure reaches a value about twice the average pressure on the projected area of the bearing (Refer to Figure 10.22).

The permissible unit pressure (PSI) is a function of the static and dynamic forces and the dimensions of the bearing (projected area). Industry guidelines for continuously loaded bearings range from 50 to 300 PSI. Pressures exceeding 300 PSI can potentially result in breakdown of the oil film and subsequent metal to metal contact.

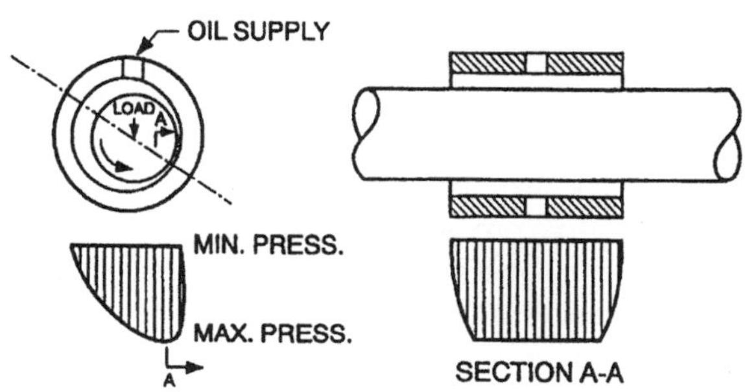

THE AVERAGE UNIT PRESSURE ON THE PROJECTED AREA (BEARING LENGTH X DIAMETER) IS GIVEN BY THE EQUATION

$$P = \frac{6ZVd}{2C^2} K'$$

Z = ABSOLUTE VISCOSITY OF LUBRICANT, CP
V = JOURNAL SURFACE VELOCITY, FPS
d = JOURNAL DIAMETER, INCHES
C = DIAMETRAL CLEARANCE BETWEEN JOURNAL AND BEARING
K′ = FACTOR DEPENDING ON BEARING CONSTRUCTION AND RATIO OF LENGTH TO JOURNAL DIAMETER

Figure 10.22 Bearing pressure distribution

Balancing devices

Single suction and multi-stage pumps require hydraulic balancing to limit thrust; whereas, double suction impellers are theoretically in hydraulic balance (Refer to Figure 10.23).

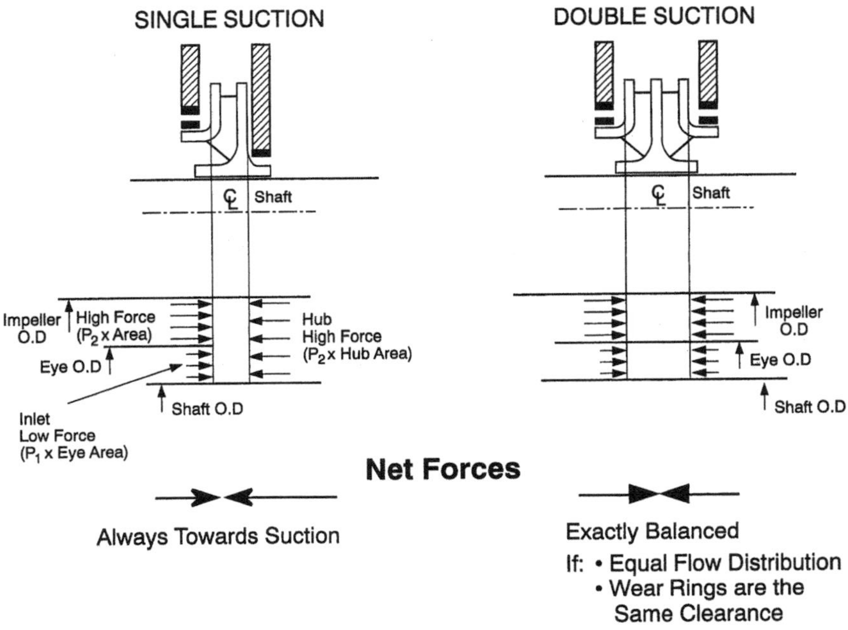

Figure 10.23 Impeller force

The thrust on a single suction impeller is always in the direction of suction. When this type of impeller is installed in a single stage overhung pump, holes are usually provided in the impeller to help maintain the stuffing box pressure in balance with the suction pressure of the pump. In some overhung pump designs pumping vanes are machined on the back hub of the impeller to reduce thrust and evacuate the area between the impeller and the stuffing box to allow flow from the stuffing box into the pump.

To minimize the impeller thrust of a multi-stage pump, two methods are used, either opposed impeller design or a hydraulic balance device. (Refer to Figure 10.24).

The claimed advantage for using opposed impeller arrangements for multi-stage pumps is its inherent zero thrust. However, there is some residual thrust resulting from leakage across bushings throughout the length of the shaft. A correctly sized thrust bearing is still required to handle the thrust load and the added thrust load resulting from

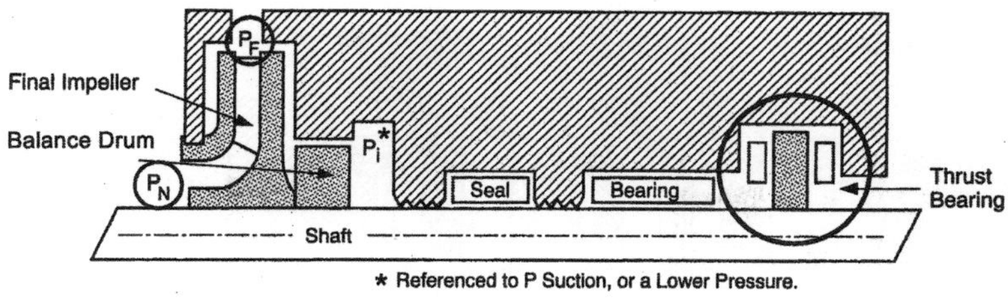

Total Impeller Thrust (LB) = Σ Individual Impeller Thrust
Balance Drum Thrust (LB) = (P_F - P_I) x (Balance Drum Area)
Thrust Bearing Load (LB) = Total Impeller Thrust – Balance Drum Thrust

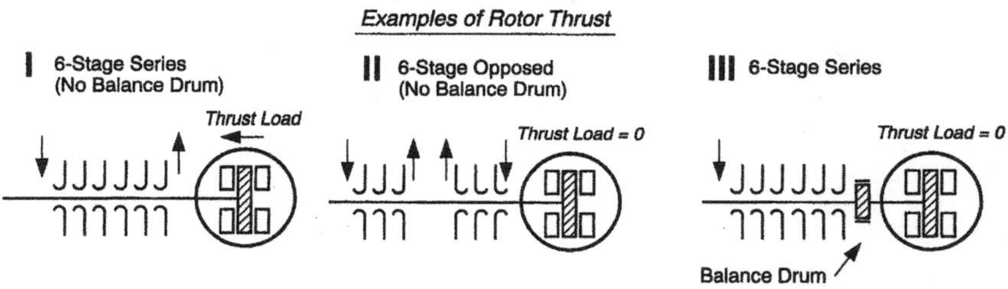

Figure 10.24 Pump thrust force

potential increased clearance of these bushings. In addition, for single stage double suction pumps, process piping design can unbalance the thrust load if turbulence is created in the suction piping as a result of unequal flow distribution to each impeller. As a rule of thumb all pumps utilizing double suction impellers should have 5–10 straight pipe diameters up stream of the suction flange.

The balance drum in Figure 10.25 is the conventional method for minimizing hydraulic axial thrust in multi-stage pumps with tandem impeller arrangements. It is not a self compensating device and thrust will increase as wear takes place and clearances open. To avoid having the rotor move back and forth in a 'shuttle' fashion, it is good practice to intentionally design the balance drum for some residual thrust. This is accomplished by sizing the drum diameter such that there is always some thrust force either in the direction of suction or discharge. The cavity behind the balance drum is routed to suction and maintained at a slight pressure above suction pressure due to the pressure drop in the balance line. Refer to Figure 10.26 for thrust bearing sizing criteria.

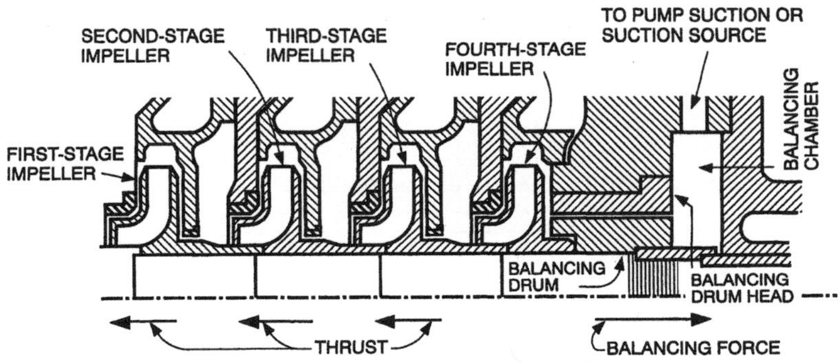

Figure 10.25 Thrust balancing drum

Thrust bearing sizing criteria

Thrust bearing is sized for:

- Net thrust load from impellers and leakage plus added thrust load resulting from twice the Increased clearance of the balance drum
- Thrust load from coupling

Figure 10.26 Thrust bearing sizing criteria

Shaft and key stress

Whenever a Root Cause Failure Analysis is required for a Shaft and/or Keyway change or the Equipment must operate under an increased power of greater than 25%, Shaft and Key Stresses should be checked. The Limiting Shaft and/or Key Stress is always the Shear Stress, which is the Stress produced by the Load (Torque) of the Equipment. Figure 10.27 presents these facts.

Always check shaft & key stress when:

- Shaft end and/or key stress damage is observed
- Equipment is being modified (upgraded) to transmit additional power

Figure 10.27

Shaft Shear Stress is calculated first and is a function of the Transmitted Torque (Proportional to Horsepower/Shaft Speed) and Shaft Diameter to the 3rd power. This relationship is shown in Figure 10.28.

Shaft shear stress

I. The relationship: $\text{Shear Stress} = \dfrac{16T}{\pi (D)^3}$

Where: $T = \text{torque} = \dfrac{\text{shaft horsepower}}{\text{shaft speed}} \times 63025$ in-LBS/HP

D = shaft diameter (smallest diameter) inches

II. Guidelines: use maximum driver hp, rated speed typical limit for STD. (4140) shaft material = 8,500 PSI

Figure 10.28 Shaft shear stress

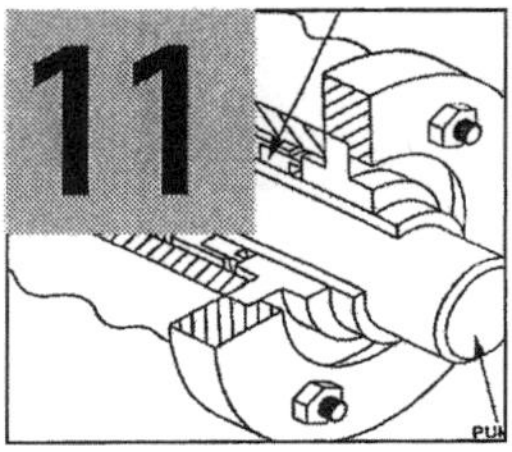

11 Mechanical seals

- Introduction
- Function of mechanical seals
- The seal system
- Controlling flush flow to the seal
- Examining some causes of seal failures
- Seal configurations
- Flush system types
- Auxiliary stuffing box and flush plans

Introduction

The part of the pump that is exposed to the atmosphere and through which passes the rotating shaft or reciprocating rod is called the stuffing box. A properly sealed stuffing box prevents the escape of pumped liquid. Mechanical seals are commonly specified for centrifugal pump applications (Refer to Figure 11.1).

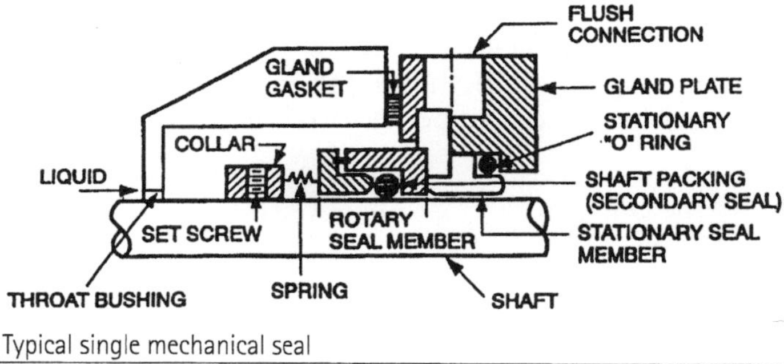

Figure 11.1 Typical single mechanical seal

Function of mechanical seals

The mechanical seal is comprised of two basic components (Refer to Figure 11.2).

Basic seal components

- Stationary member fastened to the casing
- Rotating member fastened to shaft, either direct or with shaft sleeve

Figure 11.2 Basic seal components

The mating faces of each member perform the sealing. The mating surface of each component is highly polished and they are held in contact with a spring or bellows which results in a net face loading closure force (Refer to Figure 11.1).

In order to meet the objective of seal design (prevent fluid escape to the atmosphere), additional seals are required. These seals are either 'O' rings, gaskets or packing (Refer to Figure 11.1). For high temperature applications (above 400°) the secondary seal is usually 'Graphoil' or 'Kalrez' material in a 'U' or chevron configuration. An attractive alternative is to eliminate the secondary seal entirely by using a bellows seal since the bellows replaces the springs and forms a leak tight element thus eliminating the requirement for a secondary seal (Refer to Figure 11.3).

To achieve satisfactory seal performance for extended periods of time, proper lubrication and cooling is required. The lubricant, usually the pumped product, is injected into the seal chamber and a small amount passes through the interface of the mating surfaces. Therefore, it can be stated that all seals leak and the amount of leakage depends on the pressure drop across the faces. This performance can be considered like flow through an equivalent orifice (Refer to Figure 11.4).

The amount of heat generated at the seal face is a function of the face loading and friction coefficient, which is related to material selection and lubrication. Figure 11.5 shows the equation for calculating the amount of heat which needs to be removed by the flush liquid.

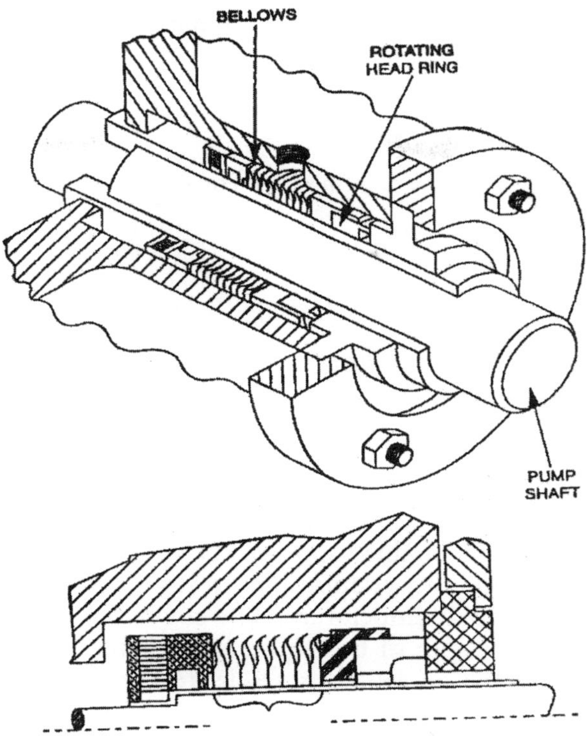

Figure 11.3 Metal bellows seal

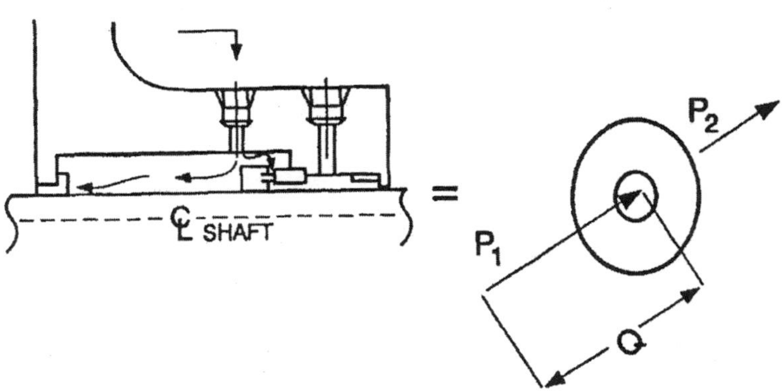

Figure 11.4 Equivalent orifice flow across seal faces

Heat generated by mechanical seal

$$Q = 500 \bullet S.G. \bullet Q_{INJ} \bullet C_p \bullet \Delta T$$

where: Q = heat load (BTU/HR)

S.G. = specific gravity of injection liquid

C_p= specific heat of injection liquid $\left(\frac{BTU}{LB-°F}\right)$

ΔT = temperature rise of injection liquid (°F)

Q_{INJ}= injection liquid flow rate (G.P.M.)

Figure 11.5 Heat generated by mechanical seal

As the lubricant flows across the interface, it is prone to vaporization. The initiation point of this vaporization is dependent upon the flush liquid pressure and its relationship to the margin of liquid vapor pressure at the liquid temperature. The closer the liquid flush pressure is to the vapor pressure of the liquid at the temperature of the liquid, the sooner vaporization will occur (Refer to Figure 11.6).

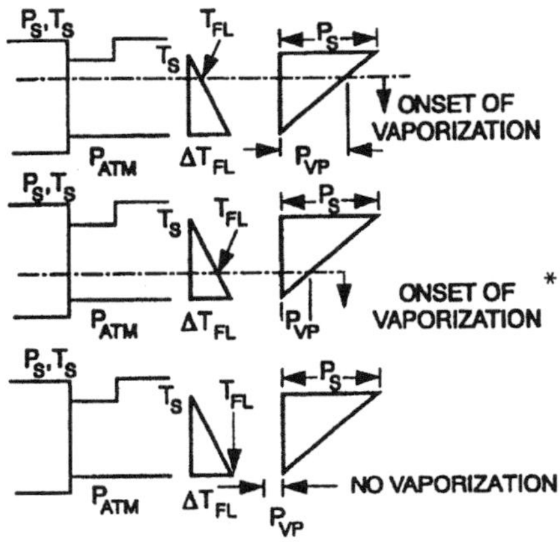

*This represents the design case, vaporization approximately 3/4 down the faces

Figure 11.6 Typical seal face pressure temperature relationship to vaporization

The seal system

To assure reliable trouble free operation for extended periods of time, the seal must operate in a properly controlled environment. This

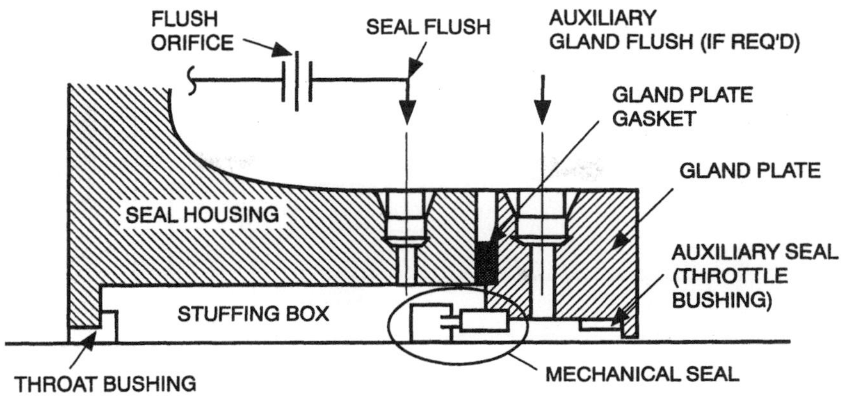

Figure 11.7 Simple seal system

requires that the seal be installed correctly so that the seal faces maintain perfect contact and alignment and that proper lubrication and cooling be provided. A typical seal system for a simple single mechanical seal is comprised of the seal, stuffing box throat bushing, liquid flush system, auxiliary seal and auxiliary flush or barrier fluid (when required) (Refer to Figure 11.7).

The purpose of the seal is to prevent leakage of pumped product from escaping to the atmosphere. The liquid flush (normally pumped product from the discharge) is injected into the seal chamber to provide lubrication and cooling. An auxiliary seal is sometimes fitted to the gland plate on the atmospheric side of the seal chamber. Its purpose is to create a secondary containment chamber when handling flammable or toxic fluids which would be considered a safety hazard to personnel if they were to leak to atmosphere. A liquid (non-toxic) flush or barrier fluid, complete with a liquid reservoir and appropriate alarm devices can be used to assure toxic fluid does not escape to the atmosphere.

Controlling flush flow to the seal

The simple seal system shown in Figure 11.8 incorporates an orifice in the flush line from the pump discharge to the mechanical seal. Its purpose is to limit the injection flow rate to the seal and to control pressure in the seal chamber. A minimum bore diameter or 1/8" is normally specified (to minimize potential of blockage) and the orifice can either be installed between flanges or in an orifice nipple.

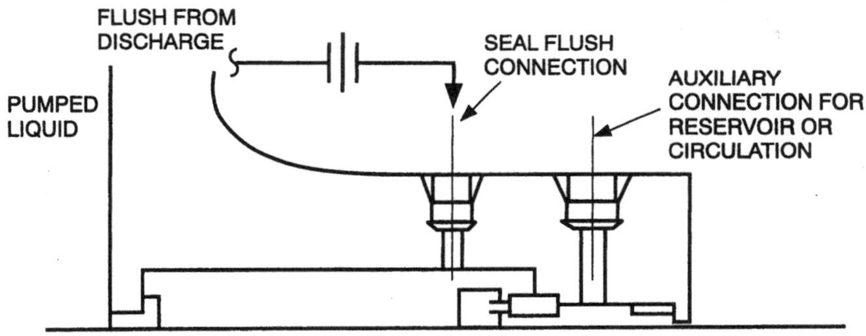

Figure 11.8 Seal flow control orifice

Examining some causes of seal failures

An indication of some causes of seal failures can be obtained while the seal is operating. When you consider the seal as an equivalent orifice, examination of 'tell tale' symptoms can indicate potential failure causes for which corrective action can be implemented or at least can provide direction of subsequent failure analysis (Refer to Figure 11.9). It should be noted that improper application, installation, and/or manufacturing errors can also result in mechanical seal failures.

Comments	Possible causes	Comments/recommendations
■ Seal squeal during operation	Insufficient amount of liquid to lubricate seal faces	Flush line may need to be enlarged and/or orifice size may need to be increased
■ Carbon dust accumulating on outside of seal area	Insufficient amount of liquid to lubricate seal faces	See above
	Liquid film vaporizing/ flashing between seal faces	Pressure in seal chamber may be too low for seal type
■ Seal spits and sputters in operation (popping)	Product vaporizing/flashing across seal faces	Corrective action is to provide proper liquid environment of the product at all times
		1. Increase seal chamber pressure if it can be achieved within operating parameters (maintain at a minimum of 25 Psig above suction pressure)

	2. Check for proper seal balance with manufacturer 3. Change seal design to one not requiring as much product temperature margin (ΔT) 4. Seal flush line and/or orifice may have to be enlarged 5. Increase cooling of seal faces Note: A review of seal balance requires accurate measurement of seal chamber pressure, temperature and product sample for vapor pressure determination

Figure 11.9 Possible causes of seal failure

Seal configurations

Mechanical seals are the predominant type of seals used today in centrifugal pumps. They are available in a variety of configurations, depending upon the application service conditions and/or the User's preference (Refer to Figures 11.10 to 11.13 for the most common arrangements used in refinery and petrochemical applications).

Single mechanical seal applications

Single mechanical seals (Refer to Figure 11.10) are the most widely used seal configuration and should be used in any application where the liquid- is non-toxic and non-flammable. As mentioned earlier in this section, many single mechanical seals are used with flammable and even toxic liquids and only rely on the auxiliary seal throttle bushing to prevent leakage to atmosphere. Since the throttle bushing does not positively contain leakage, State and Federal environmental regulations now require use of a tandem or double seal for these applications. In some plants, a dynamic type throttle bushing ('Impro' or equal) is used to virtually eliminate leakage of the pumped fluid to atmosphere in the event of a mechanical seal failure.

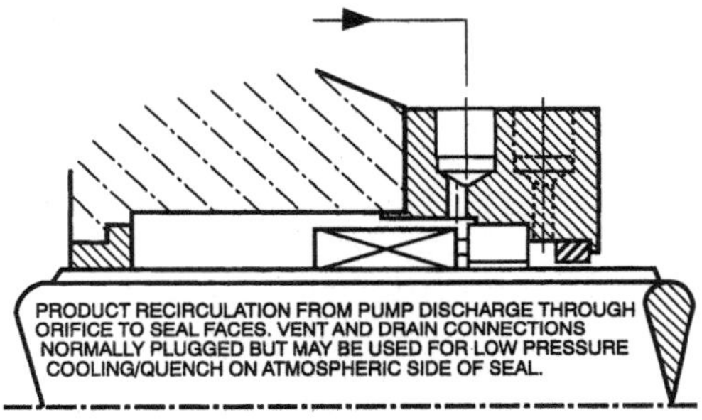

APPLICATIONS: NON HYDROCARBON, HYDROCARBON LIQUIDS. SPECIAL FEATURES INCORPORATED DEPENDING ON CHEMICAL CONTAMINANTS

Figure 11.10 Single mechanical seal

Tandem mechanical seal applications

Tandem mechanical seals (Refer to Figure 11.11) are used in applications where the pumped fluid is toxic and/or flammable. They consist of two (2) mechanical seals (primary and back-up). The primary seal is flushed by any selected seal flush plan. The back-up seal is provided with a flush system incorporating a safe, low flash point liquid. A pressure alarm is provided to actuate on increasing stuffing box pressure between the primary and back-up seal thus indicating a primary seal failure. Since the pumped product now occupies the volume between the seals, failure of the back-up seal will result in leakage of the pumped fluid to atmosphere. In essence any time a tandem seal in alarm, it is actually a single seal and should be shut down immediately to assure that the toxic and/or flammable liquid does not leak to atmosphere.

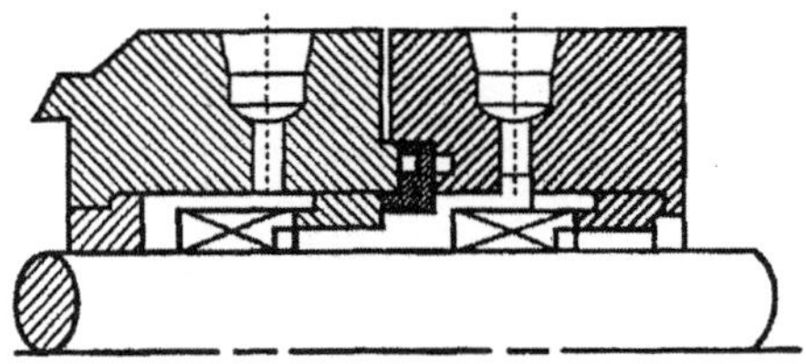

APPLICATIONS: TOXIC, EXPLOSIVE HAZARD FROM LEAKAGE, CRYOGENIC LIQUIDS

Figure 11.11 Tandem seal arrangement

Double mechanical seal applications

Double mechanical seals (Refer to Figure 11.12) are used in applications where the pumped fluid is flammable or toxic and leakage to atmosphere cannot be tolerated under any circumstances. Typical process applications for double seals are H_2S service, Hydrofluoric acid alkylation services or sulfuric acid services.

Leakage of the pumped fluid to the atmosphere is positively prevented by providing a seal system, whose liquid is compatible with the pumped liquid, that continuously provides a safe barrier liquid at a pressure higher than the pumped fluid. The seals are usually identical in design with the exception that one seal incorporates a pumping ring to provide a continuous flow of liquid to cool the seals. Typical double seal system components are: reservoir, cooler, pressure switch and control valve.

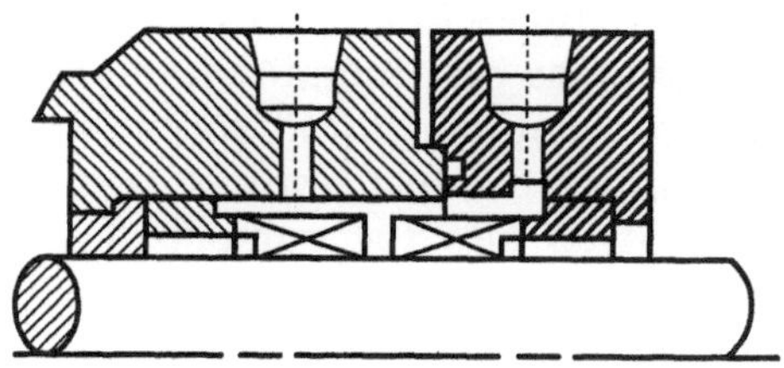

APPLICATIONS: SIMILAR TO THESE USED FOR TANDEM SEALS

Figure 11.12 Double mechanical seal

Liquid/Gas tandem mechanical seal applications

In this configuration (Refer to Figure 11.13) a conventional single liquid mechanical seal is used as the primary and a gas seal (non-contacting faces) that can temporarily act as a liquid seal in the event of primary seal failure serves as the back-up seal. This seal configuration is used in low specific gravity applications where the pumped fluid is easily vaporized. Using a gas seal as the back-up has the advantage of eliminating the vessel, cooler and pumping ring necessary for conventional tandem liquid seals.

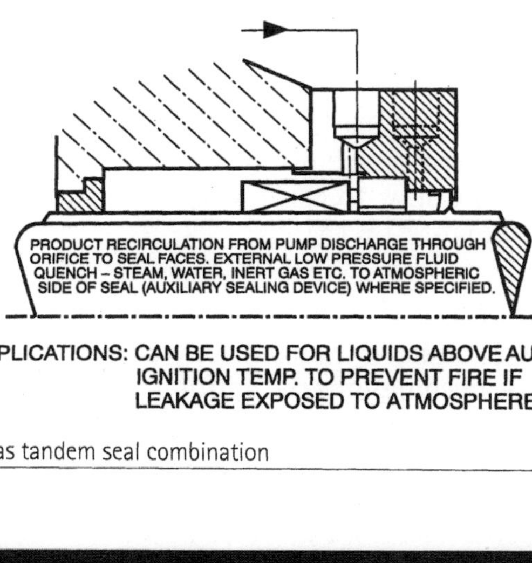

APPLICATIONS: CAN BE USED FOR LIQUIDS ABOVE AUTO IGNITION TEMP. TO PREVENT FIRE IF LEAKAGE EXPOSED TO ATMOSPHERE

Figure 11.13 Liquid/gas tandem seal combination

This application is well proven and has been used successfully for natural gas liquids, propane, ethylene, ethane and butane pump applications.

Double gas seal applications

Before leaving this subject, a relatively new application utilizes two (2) gas seals in a double seal configuration and uses N_2 or air as a buffer maintained at a higher pressure than the pumped fluid to positively prevent the leakage of pumped fluid to atmosphere. This configuration, like the tandem liquid/gas seal mentioned above eliminates the seal system required in a conventional liquid double seal arrangement. Note however, that the pumped product must be compatible with the small amount of gas introduced into the pumped fluid. This configuration cannot be used in recycle (closed loop) services.

An excellent resource for additional information covering design, selection and testing criteria, is API Standard 682.

Flush system types

Providing the proper environment for the seal is a key factor in achieving satisfactory seal operation. In conjunction with defining the seal design and materials of construction, it is necessary to decide which type of seal flush system will be selected to lubricate and cool the seal faces. The API industry has developed various systems to accommodate the requirements of almost every possible sealing arrangement (Refer to Figure 11.14 and 11.14A).

Clean product systems (Plan 11)

In this plan, product is routed from the pump discharge to the seal chamber for lubricating and cooling the seal faces. It will also vent air and/or vapors from the chamber as it passes to the pump suction through the throat bushing. NOTE: For single stage pumps without back hub pumping vanes, it is necessary to have holes in the impeller that reduce the pressure behind the impeller below the discharge pressure to allow flush flow to exit the stuffing box through the throat bushing. (Refer to Figure 11.15).

Clean product flush (Plan 13)

In this plan, the product is routed from behind the pump impeller, through the throat bushing into the stuffing box and out of the stuffing box, through an orifice back to the suction (Refer to Figure 11.16)

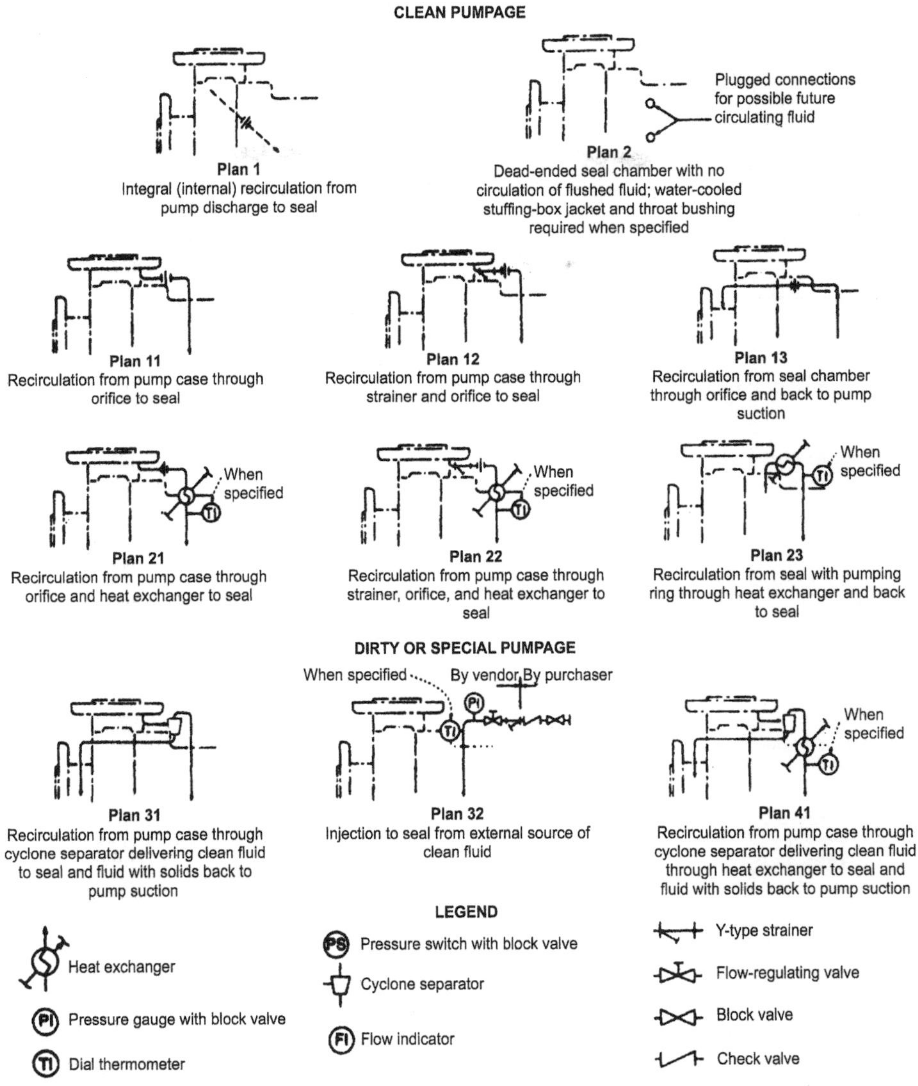

Figure 11.14 All flush plans reprinted with the permission of The American Petroleum Institute

This plan is used primarily in vertical pump applications because it provides positive venting of the stuffing box. It is also used in single stage pump applications that do not employ pumping vanes or holes in the impellers.

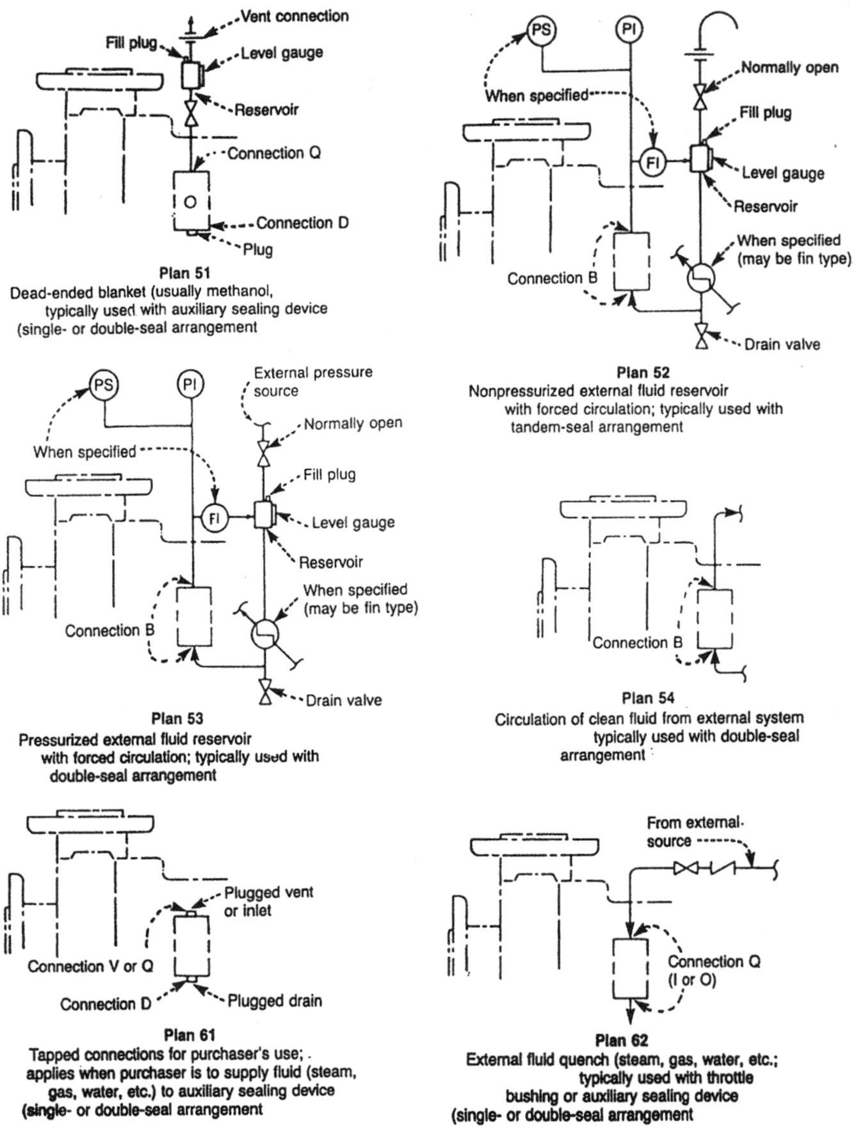

Figure 11.14A API flush plans continued

Dirty product system (Plan 31)

Liquid product from the pump discharge is routed to the seal chamber through a cyclone separator which is selected to optimize removal of solids across an individual pump stage (Refer to Figure 11.17 and to Figure 11.18 for guidelines covering the use of cyclones).

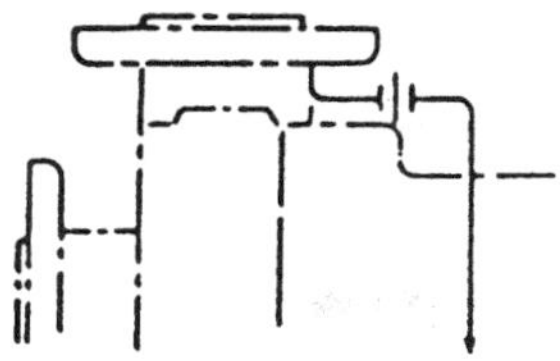

Plan 11
Recirculation from pump case through orifice to seal

Figure 11.15 API flush plan 11

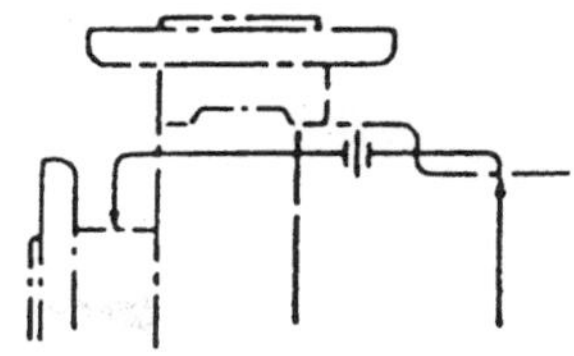

Plan 13
Recirculation from seal chamber through orifice and back to pump suction

Figure 11.16 API flush plan 13

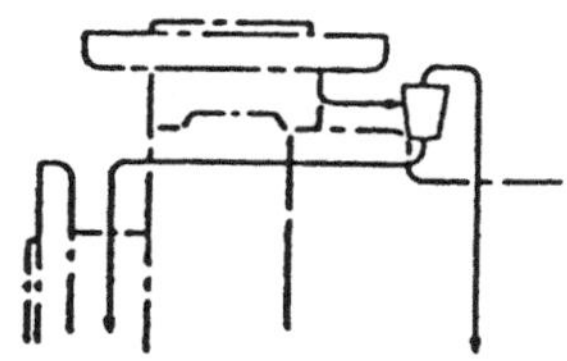

Plan 31
Recirculation from pump case through cyclone separator delivering clean fluid to seal and fluid with solids back to pump suction

Figure 11.17 API flush plan 31

Guidelines for use of cyclones

- Do not use cyclones when differential pressure is less that 25 Psi
- Consider using orifice when pressure differential exceeds cyclone design differential
- Solids to be removed should have density at least twice that of the fluid
- Efficiency of separation is reduced as differential pressure across cyclone varies from design differential
- Separation efficiency drops as particle size decreases

Figure 11.18 Guidelines for use of cyclones

When a clean, cool seal flush liquid is mandated for reasons of preventing solids accumulation in the seal chamber, an external liquid flush system (Plan 32) is used. The fluid pressure is higher than that behind the impeller so that flow is always towards the pump suction, thus preventing back flow of dirty product into the seal chamber (Refer

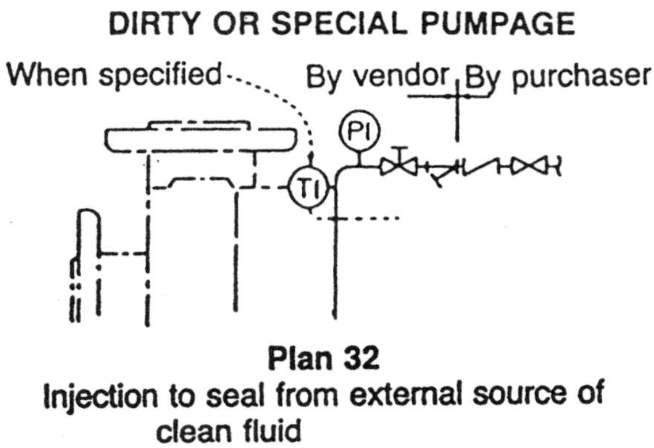

Figure 11.19 API flush plan 32

to Figure 11.19). When using Flush Plan 32, it is important to confirm that the flush fluid will not vaporize in the stuffing box.

High temperature product flush system (plan 23)

This flush plan is desirable when it is necessary to maintain the required margin between liquid vapor pressure (at seal chamber temperature) and seal chamber pressure. The feature about this plan is that the cooler only removes heat generated by the seal faces plus the heat soak through the shaft from the process. A throat bushing is installed in the seal chamber to isolate the product in this area from that in the impeller area of the pump. A circulating device (pumping ring) is mounted on the seal which circulates liquid in the seal chamber through a cooler and back to the seal chamber. It is more efficient than Plans 21 and 22 which incorporate a cooler to continuously cool the flush from the discharge. These plans are simply Flush Plan 1 1 with the addition of a cooler. Flush Plans 21, 22 or 23 should be used when the temperature of the pumped fluid is above 400°F and a bellows seal is not applied. It is important to confirm that the pumped fluid is clean when using Flush Plan 23 since there is not a constant external flush. This plan is typically used for boiler feed pump applications (Refer to Figure 11.20).

Low temperature product flush/buffer system (plan 52)

This system is well suited for low temperature applications such as ethylene, propylene and other low temperature liquids which are susceptible to forming ice on the seal faces when the atmospheric side of the seal is exposed to the atmosphere; thus separating the faces and resulting in excessive seal leakage. This plan consists of a tandem (dual) seal with a buffer liquid between them. A seal pot containing the buffer liquid (usually methanol – a drying agent) is vented to a lower pressure vent system. The seal pot system is usually equipped with a pressure

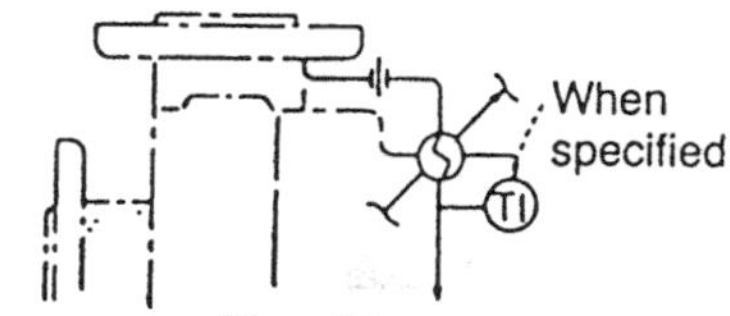

Plan 21
Recirculation from pump case through orifice and heat exchanger to seal

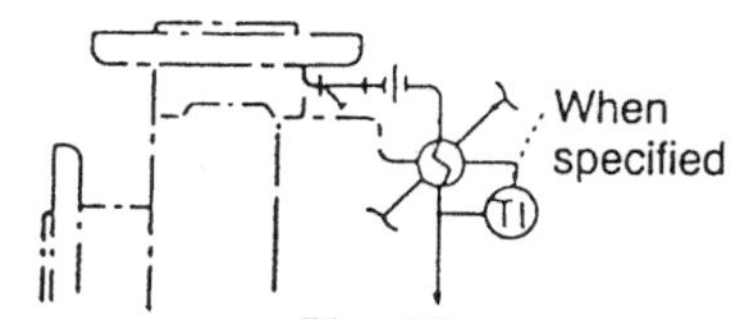

Plan 22
Recirculation from pump case through strainer, orifice, and heat exchanger to seal

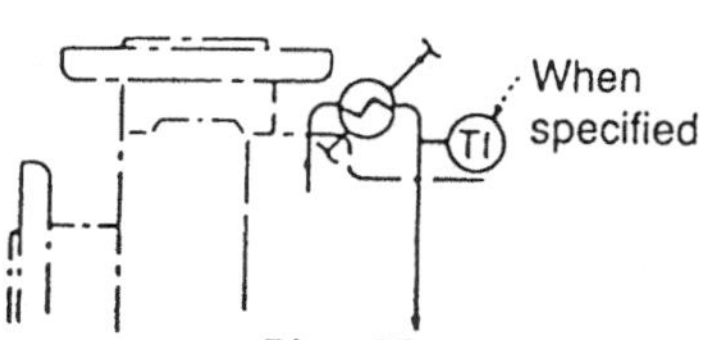

Plan 23
Recirculation from seal with pumping ring through heat exchanger and back to seal

Figure 11.20 API high temperature flush plans 21, 22, 23

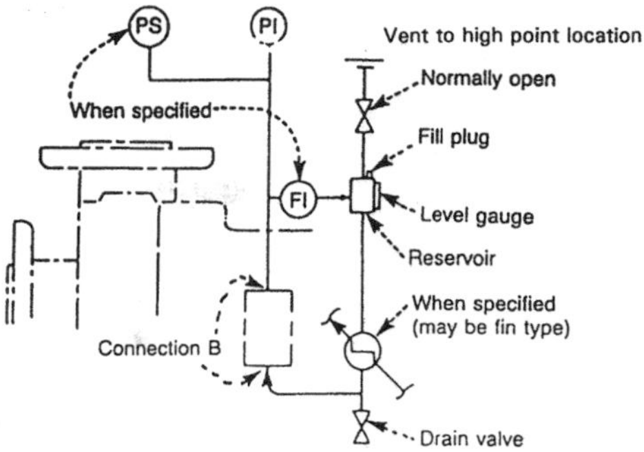

Plan 52
Nonpressurized external fluid reservoir with forced circulation; typically used with tandem-seal arrangement

Figure 11.21A plan 52

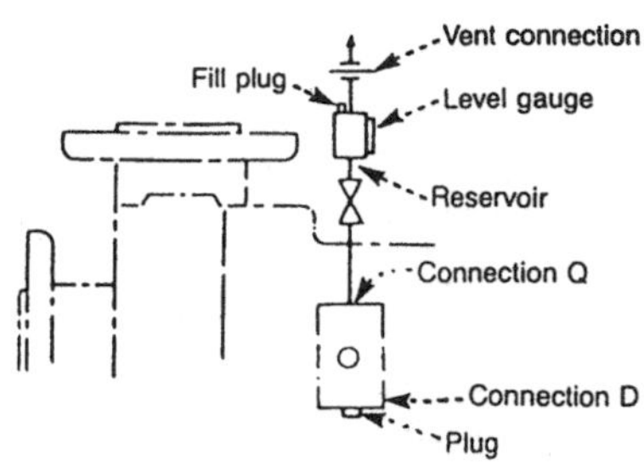

Plan 51
Dead-ended blanket (usually methanol, see Note 3); typically used with auxiliary sealing device (single- or double-seal arrangement in Figure D-1)

Figure 11.21B plan 51

switch to sound an alarm if the inner seal product leakage cannot be adequately carried away through the orifice vent system (Refer to Figure 11.21A). When the alarm sounds, the pump should be shut down as soon as possible since the back-up seal is now functioning as the primary seal and will leak the pumped fluid to atmosphere if it fails.

Seal flush Pan 51 may also be used when leakage to atmosphere cannot be tolerated. Plan 51 incorporates a dead ended system design and is shown in Figure 11.21B.

Toxic or flammable product flush system (plan 53)

This system is used when leakage to the atmosphere cannot be tolerated (see Figure 11.22A). It consists of a dual seal arrangement with a barrier liquid between them. A seal pot contains the barrier liquid at a

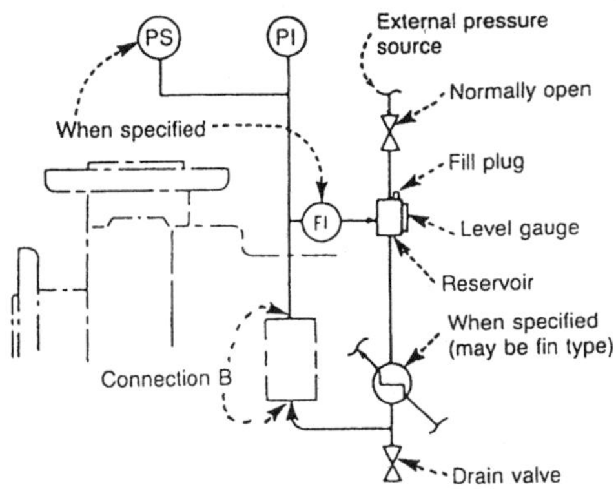

Plan 53
Pressurized external fluid reservoir (see Note 3) with forced circulation; typically used with double-seal arrangement in Figure D-1

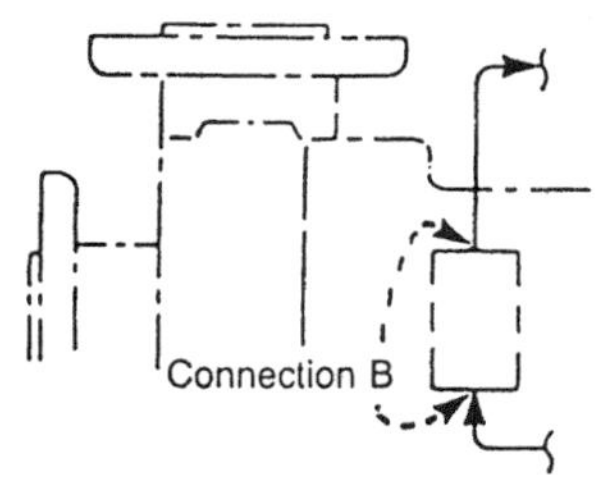

Plan 54
Circulation of clean fluid from external system typically used with double-seal arrangement

Figure 11.22 A & B plans 53 and 54

pressure higher than seal chamber pressure (usually 20–25 PSI). Inner seal leakage will always be barrier liquid leakage into the product, resulting in some product contamination. The barrier liquid should be selected on the basis of its compatibility with the product. An internal pumping device (pumping ring) is used to circulate the barrier liquid into and out of the seal chamber through the seal pot. The integrity of the system always needs to be monitored to assure that seal pot pressure level is maintained with barrier liquid.

Plan 54 is also a dual system which utilizes a pressurized barrier liquid from an external reservoir or system to supply clean cool liquid to the seal chamber. As described in the previous plan, the barrier liquid pressure level is higher than the seal chamber pressure (usually 20–25 PSI) so that inner seal leakage is always into the pump. With this plan, it is also necessary to consider the compatibility of the barrier liquid and the pumped product. This system is considered to be one the most reliable systems available. However, it is more complex and more costly than other systems (Refer to Figure 11.22B).

Auxiliary stuffing box and flush plans

As mentioned previously, the auxiliary stuffing box can be used as an auxiliary sealing device in the event of seal failure. (Refer to Figure 11.23). It is important to note that the auxiliary stuffing box contains a restricted flow seal (packing or close fitting throttle bushing) and does not positively contain the pumped fluid. Therefore, the auxiliary

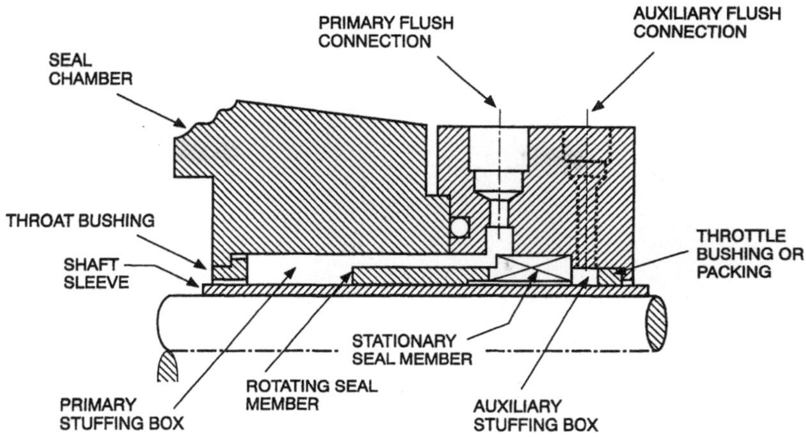

Figure 11.23 Auxiliary stuffing box

stuffing box seal device is for emergency containment of the pumped fluid only. The pump should be shut down immediately in the event of leakage observed from the auxiliary stuffing box if a quench is not supplied. It is always a good practice to require that the auxiliary stuffing box drain connection, which will come plugged, be piped to a drain system that meets environmental standards.

When the pumped fluid can vaporize and form hard deposits, the auxiliary stuffing box is used to contain a quench fluid that will dissolve (wash away) the hard deposits at the exit of the seal faces, thus eliminating seal face wear.

A typical refinery quench application is the use of low pressure (50 PSI) steam in and out of the auxiliary stuffing box to dissolve coke deposits on the seal face. This application also has the added advantage of keeping the standby pump seal warm which prevents thermal expansion problems in start-up. This arrangement uses a throttle bushing as the external seal.

Water is also used as an auxiliary stuffing box flush in caustic applications where solid deposits need to be flushed from the seal face. When a water flush is used, two (2) rows of packing are usually provided in the auxiliary stuffing box as an external seal to minimize external water leakage.

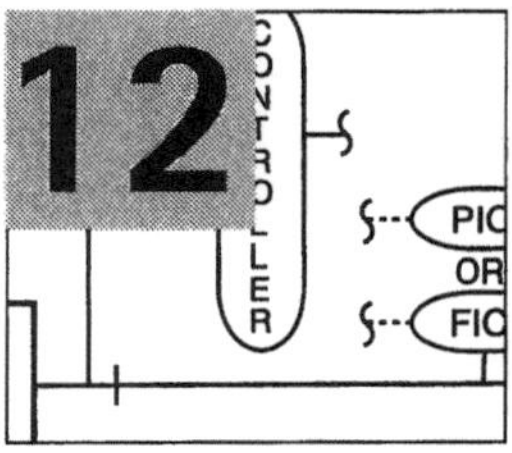

12 Pump control and protection

- Introduction
- Objectives of the operating company
- Adjusting the head produced by a centrifugal pump
- Varying process flow of a positive displacement pump
- Multiple pump applications
- The affinity laws
- Protecting the pump
- Pump protection systems

Introduction

Thus far, we have discussed hydraulic and mechanical design aspects of pumps commonly used in the petroleum, petrochemical and gas services industries. We have seen that there are two (2) basic types of pump characteristics – positive displacement and dynamic. In order to meet the end user's objectives, these characteristics must be modified by utilizing reliable control and protection schemes. We will now focus the discussion on control and protection systems for achieving site maximum safety and reliability.

Objectives of the operating company

During normal operation, the prime objective of the operating company is to maintain maximum reliability and maximum product

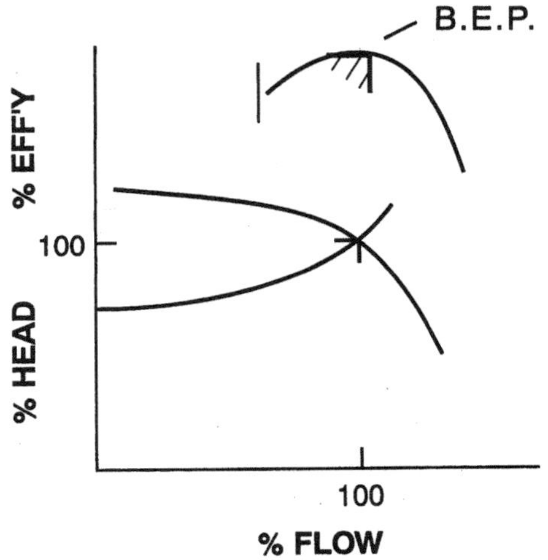

- Maximum reliability (on stream time)
- Maximum product throughput
- Minimum operating costs

Figure 12.1 The end user's objectives

throughput at minimum operating (maximum pump efficiency) cost (refer to Figure 12.1).

To meet the operating company objectives, a particular pump is selected based on the optimum process flow rate and the required system head (energy). Since the pump characteristics will change with wear, erosion or fouling and the system characteristics will vary, a reliable control and protection system must be selected to continuously meet the objectives noted in Figure 12.1 regardless of whether the selected pump is dynamic or positive displacement. Regardless of the pump type, the operating company's objectives are met by meeting process throughput requirements in the most efficient manner.

There are only two options available to vary the process throughput requirements (refer to Figure 12.2).

Methods for varying pump throughput

- Adjust the head required
- Adjust the head produced

Figure 12.2 Methods for varying pump throughput

Before proceeding, it is helpful to introduce the concept of an equivalent vessel. Another way to state the operating company's objectives is to state that they want to process all the throughput produced. If we visualize any process system as consisting of a vessel into which the produced product flows, the objective then is to remove (pump) the throughput from the vessel at the same rate of flow entering the vessel. If this is accomplished, the following parameters will remain constant in the vessel:

- Flow in and out
- Level
- Pressure

Any point in any process can be thought as an "equivalent vessel". By controlling either constant flow, level or pressure, the operating company's objective will be realized (refer to Figure 12.3).

Therefore the objective of any pump control system will be to maintain a constant controlled variable set point (flow, level or pressure). The output of the controller can then either vary:

- Head required (the process)
- Head produced (the pump characteristics)

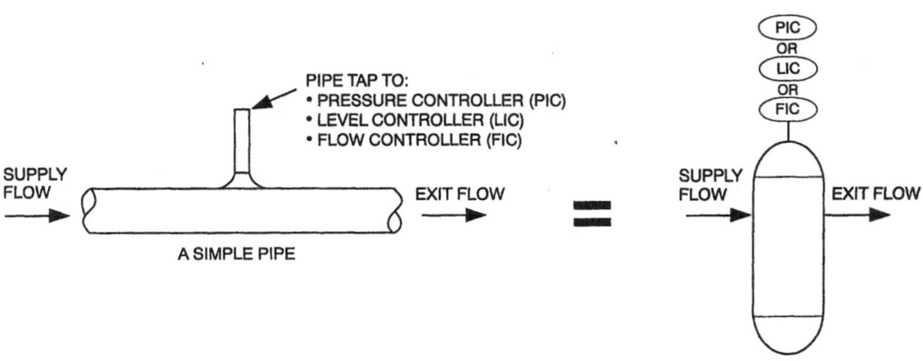

Case	Equivalent Vessel Variable
SUPPLY = EXIT FLOW	CONSTANT
SUPPLY FLOW GREATER THAN EXIT FLOW	INCREASES
SUPPLY FLOW LESS THAN EXIT FLOW	DECREASES

Figure 12.3 Reduce it to an equivalent vessel

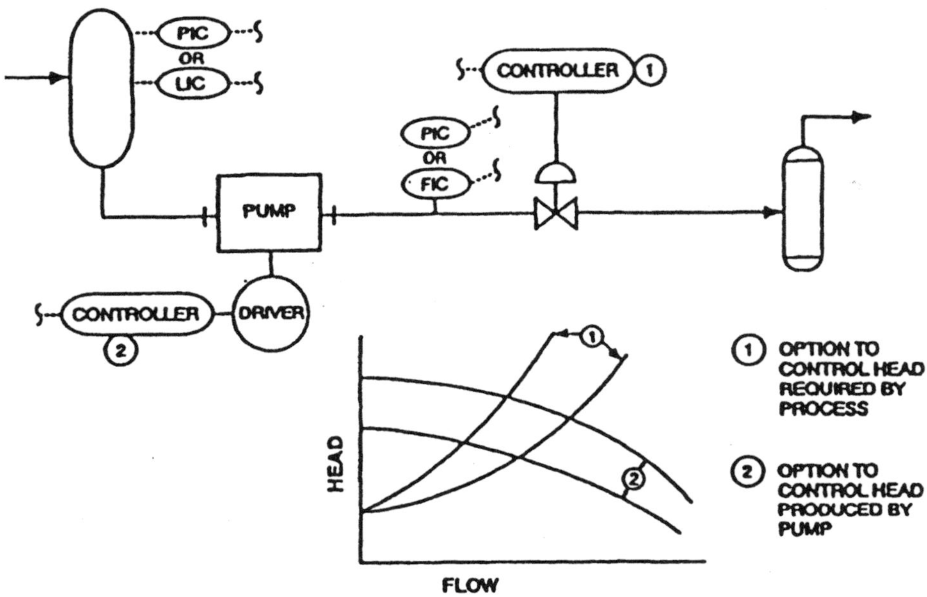

Figure 12.4 Centrifugal pump control options

These options are shown in Figure 12.4 for a centrifugal pump and Figure 12.5 for a P.D. pump.

Adjusting head (energy) required

Head required in a pumping system can be changed by adjusting the discharge system resistance using pressure control flow control or level control (refer to Figures 12.4 and 12.5).

Each of these methods results in closing a throttle valve in the discharge piping which increases the head (energy) required and reduces the flow rate. This action requires more energy (head) to overcome the increased system resistance (refer to Figure 12.6)

Throttling the discharge of a positive displacement pump increases the discharge pressure and consequently the power required to overcome the increased head required while passing the same flow rate through the pump. This scheme is not efficient and is usually replaced by a "bypass" throttling arrangement (refer to Figure 12.5).

Effects of throttling pump suction

Generally, throttling the pump suction will provide harmful effects and should be avoided (absolute pressure is reduced at the impeller inlet)

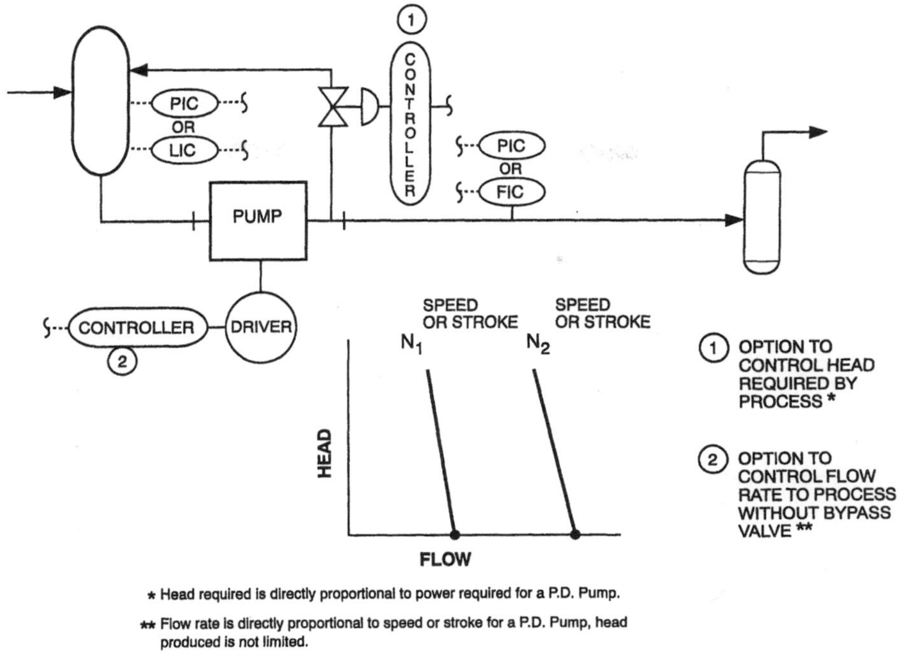

Figure 12.5 P.D. pump control options

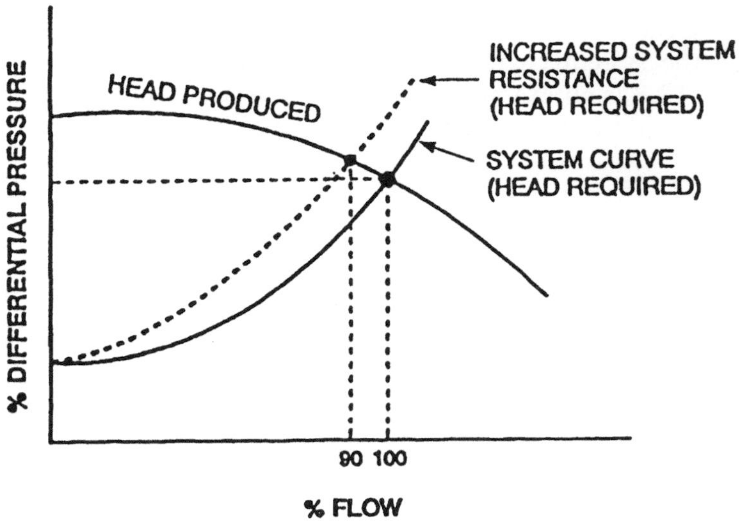

Figure 12.6 Effect of discharge throttling

except in certain cases involving series pump operation and "hot well" condensate pumps specifically designed for such services (refer to Figure 12.7 for effects of suction throttling).

Effects of suction throttling

- Cavitation
- Overheating
- Seal failure
- Bearing failure

Figure 12.7 Effects of suction throttling

Adjusting the head produced by a centrifugal pump

When dealing with a system comprised mainly of friction losses, speed control is the most effective way to vary the flow through a centrifugal pump because the efficiency remains essentially constant. Operating costs are reduced since higher efficiency results in lower power input.

With speed control the pump characteristics (head produced) vary while the system characteristics (head required) remains unchanged. Figure 12.8 is an example of decreasing head required being met by a reduction in head produced by a reduction in impeller tip speed.

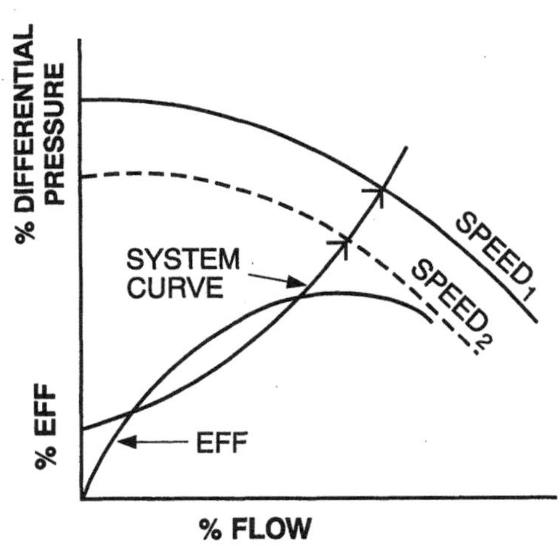

Figure 12.8 Speed control characteristics

Speed control systems most commonly used in the petroleum industry can be effected by a variety of variable speed driver systems (refer to Figure 12.9)

Common methods to vary speed of centrifugal pumps

- Steam turbine
- Gas turbine
- Diesel engine
- Variable speed electric motor
- Hydraulic coupling

Figure 12.9 Common methods to vary speed of centrifugal pumps

Varying process flow of a positive displacement pump

Based on the characteristics of a positive displacement pump being constant flow delivery versus variable energy (head), there are basically two options available for varying the flow through the pump (refer to Figure 12.10).

Flow control methods for positive displacement pumps

- Bypass control
- Speed control

Figure 12.10 Flow control methods for positive displacement pumps

P.D. pump bypass control

Bypass control is probably the least costly of the two methods and the one most commonly used in the petroleum, petrochemical and gas service industries. Excess flow not required by the process is routed back to the suction source. The controlled variable for the bypass control valve can be process flow, process pressure or possibly liquid level control. As one of these variables change, a signal is transmitted to the control valve to either open further or close (refer to Figure 12.11 for a typical P.D. pump bypass system).

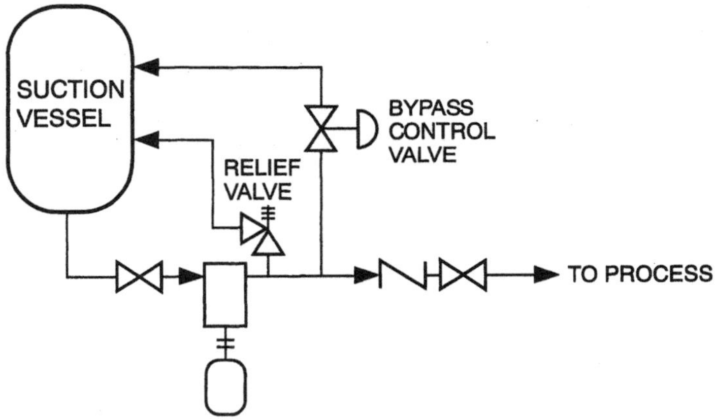

Figure 12.11 Typical P.D. pump bypass system

P.D. pump speed control

Speed change is an alternative method for varying flow through a P.D. pump and it can be accomplished with driver systems similar to those used for centrifugal pumps (refer to Figure 12.12). When applying variable speed drivers to positive displacement pumps, the potential for torsional vibrations must be considered.

Methods for varying speed of positive displacement pumps

- Steam turbine
- Diesel engine
- Variable speed electric motor
- Hydraulic coupling

Figure 12.12 Methods for varying speed of positive displacement pumps

Direct acting and metering pump control

Direct acting pumps and metering pumps (controlled volume positive displacement pumps) are subjected to variable flow requirements and the method for varying the flow through each type of pump is reviewed in Figure 12.13.

Controlling flow to direct acting and controlled volume P.D. pumps

- Direct acting pump
  Flow varies by regulating the supply of motive fluid resulting in action to either slow down or speed up pump based on control parameter
- Metering pump (controlled volume P.D. pump)
  Stroke length adjustment
  - Manual
  - Automatic by external electronic or pneumatic signal

  Speed adjustment
  - Variable speed electric motor

Figure 12.13 Controlling flow to direct acting and controlled volume P.D. pumps

Multiple pump applications

When one (1) pump cannot efficiently or mechanically meet process requirements, either parallel or series pump arrangements are required.

Parallel pump operation

When flow requirements are variable, it may be more cost effective overall to operate two pumps in parallel rather than use a single large pump. As demand drops off, one pump can be shut down, allowing the remaining pump to operate at or near its peak efficiency. Centrifugal pumps which operate most effectively in parallel are identical ones with steadily rising curves from rated flow to shutoff (refer to Figure 12.14). To obtain the overall curve for any pumps operating in parallel, add the flows from each pump at equal heads.

It is advisable to check performance of pumps to operate in parallel before attempting to do so. Frequently, the main pump has been operated much longer than the auxiliary pump and will have experienced wear and as a result, produce a different head vs flow characteristic. Figure 12.15 shows the effect on the head-capacity characteristic of two pumps in parallel if one pump is deficient in head produced. Pumps in parallel operation should be protected by a minimum flow bypass system to prevent operation at shut off (zero flow).

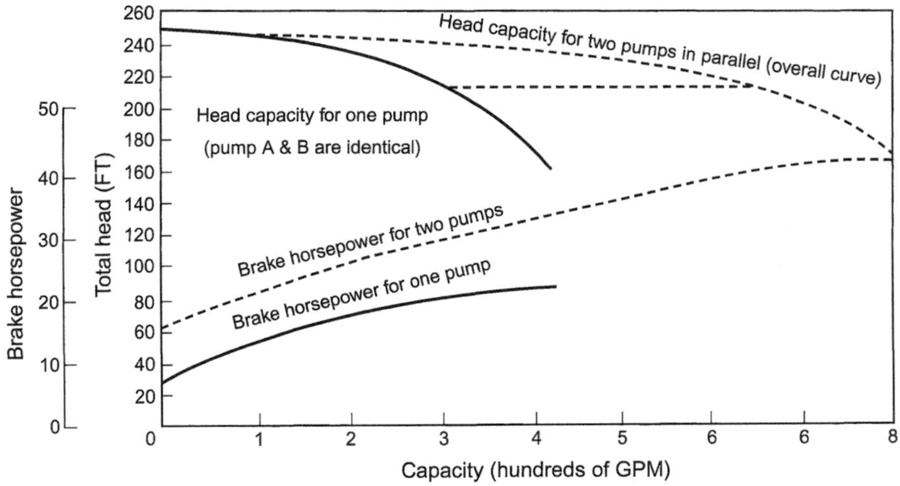

Figure 12.14 Parallel pump operation – identical pumps

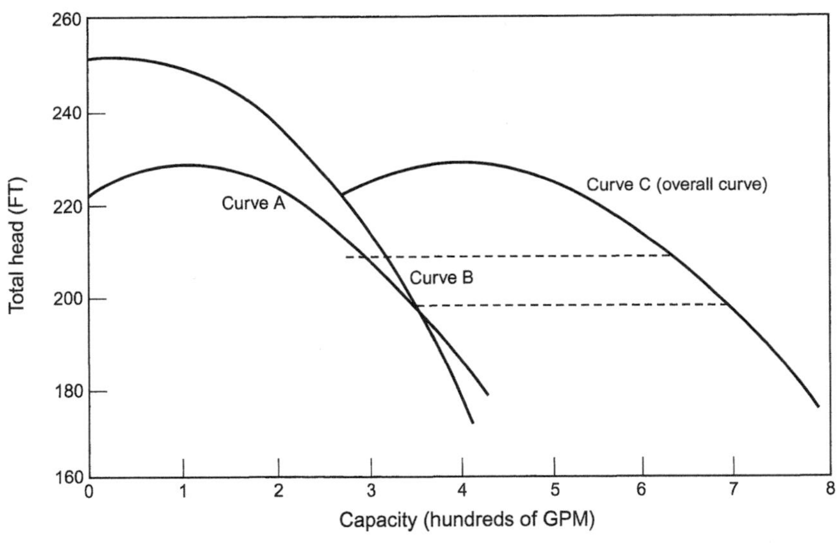

*Note that overall curve C includes part of curve B from 0 flow to 250 gpm since pump A will not pump umtil its head is equal to the head of B pump

Figure 12.15 Parallel pump operation – non identical pumps

Series pump operation

The characteristic curves for centrifugal pumps operating in series do not have to be similar for satisfactory results because the energy (head) developed by one adds to the other. To obtain the overall curve, add heads at equal flows. Generally, however, the pumps should have steadily rising head capacity curves to assure stable operation (refer to Figure 12.16).

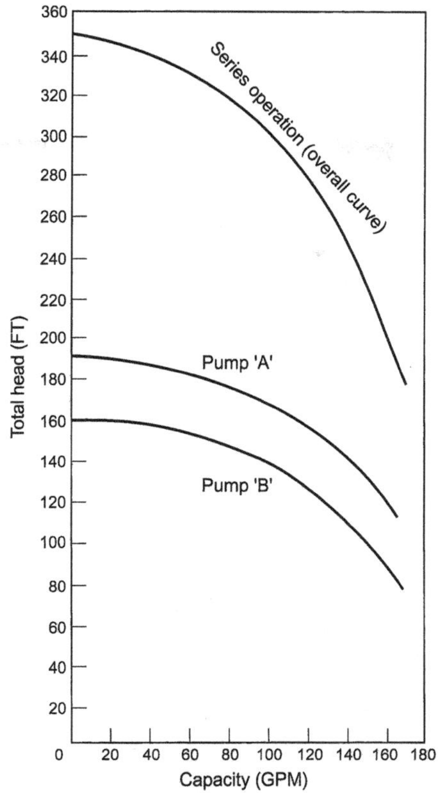

Figure 12.16 Series pump operation

It is important to consider the pressure ratings of pump cases, seals and thrust bearing capacity to assure each pump's mechanical design limits are not exceeded when operating in series.

Pumps operating in series should be protected, if necessary, by relief or pressure control valves to prevent overpressure.

The affinity laws

Let us examine the effect of speed and/or impeller diameter change on the performance of centrifugal pumps.

The affinity laws or fan laws, as they are sometimes referred to, play an important role in determining centrifugal pump performance for changes in operating conditions. They are also used for scale up purposes when performance parameters exceed that of existing pumps (refer to Figure 12.17 for affinity law relationships).

Affinity law relationships

$$\frac{Q_1}{Q_2} = \frac{N_1}{N_2} = \frac{D_1}{D_2}$$

$$\frac{H_1}{H_2} = \left(\frac{N_1}{N_2}\right)^2 = \left(\frac{D_1}{D_2}\right)^2$$

$$\frac{HP_1}{HP_1} = \left(\frac{N_1}{N_2}\right)^3 = \left(\frac{D_1}{D_2}\right)^3$$

Where: Q = Flow, GPM
N = Speed, RPM
H = Head, ft.
H_P = Horsepower
D = Impeller diameter inches

Figure 12.17 Affinity law relationships

Once a pump has been selected and the impeller diameter has been determined to deliver a defined flow rate for a required level of head (energy), the affinity laws can be used to determine what new speed or impeller diameter is required to satisfy the alternative operating conditions. A revised flow versus head (energy) versus horsepower curve can be developed and plotted from these relationships. Refer to Figure 12.18 for the example of an impeller diameter change.

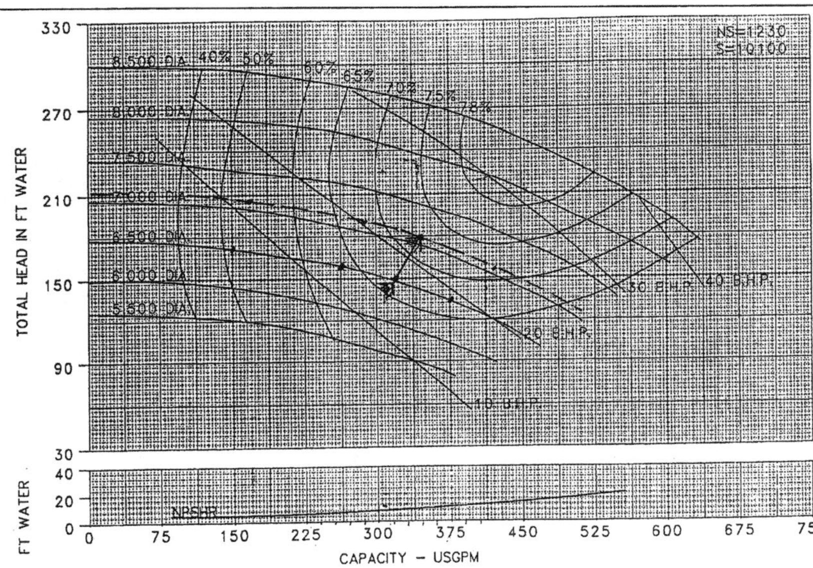

Exist flow	New flow	Exist head	New head	Exist DIA	New DIA	Speed	Exist HP	New HP	Exist EFF %	New EFF %
315	345	150	180	6.5	7.12	3550	17.6	23.2	67.5	72.5
262	287	159	190	6.5	7.12	3550	16.4	22	64	68.0
150	164	171	205	6.5	7.12	3550	13	17	51	53.5
375	411	132	158	6.5	7.12	3550	16.3	22	67.5	71.5

Figure 12.18 Performance change vs impeller diameter change (Courtesy of Union Pump Company)

The affinity law relationships for changing impeller diameter usually works pretty well for relatively small changes, on the order of 10%. When the change exceeds 10%, the relationship between the impeller and casing can change significantly to potentially alter the design configuration of the pump. It is always good practice to check the pump rating curves to determine whether the pump has been tested with that particular impeller diameter.

The variation of pump performance with changes in speed also follows the affinity laws but with a higher level of accuracy as compared to changes in impeller diameter.

When the performance is known for a pump operating at a lower speed (eg 1150 rpm) and it is desired to operate at a higher speed (eg 1750 rpm), the affinity laws can be used to calculate performance at the new speed and the converse is also true. However, it is good practice to verify the mechanical integrity of the pump coupling and driver with the vendor before proceeding with the change. The mechanical design of the process system must also be confirmed for the revised pressures and flow rates (refer to Figure 12.19)

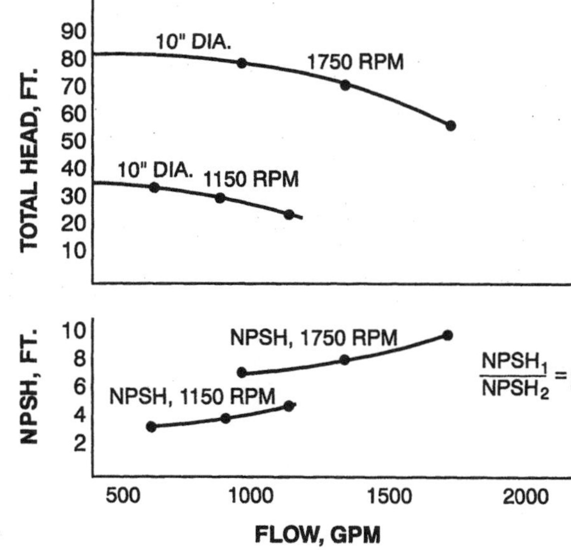

$$\frac{NPSH_1}{NPSH_2} = \left(\frac{N_1}{N_2}\right)^{1.7}$$

$$\frac{Q_1}{Q_2} = \frac{N_1}{N_2}; \quad \frac{H_1}{H_2} = \left(\frac{N_1}{N_2}\right)^2; \quad \frac{HP_1}{HP_2} = \left(\frac{N_1}{N_2}\right)^3$$

EXIST FLOW	EXIST HEAD	EXIST NPSH	EXIST HP	EXIST RPM	EXIST EFF	NEW FLOW	NEW HEAD	NEW RPM	NEW HP	NEW NPSH
612	35	3.5	9	1150	60	931	81	1750	32	7.14
875	32	4.0	10	1150	70	1331	74	1750	36	8.16
1137	26	5.0	10.5	1150	71	1730	60	1750	37	10.2

Figure 12.19 Performance change versus speed change

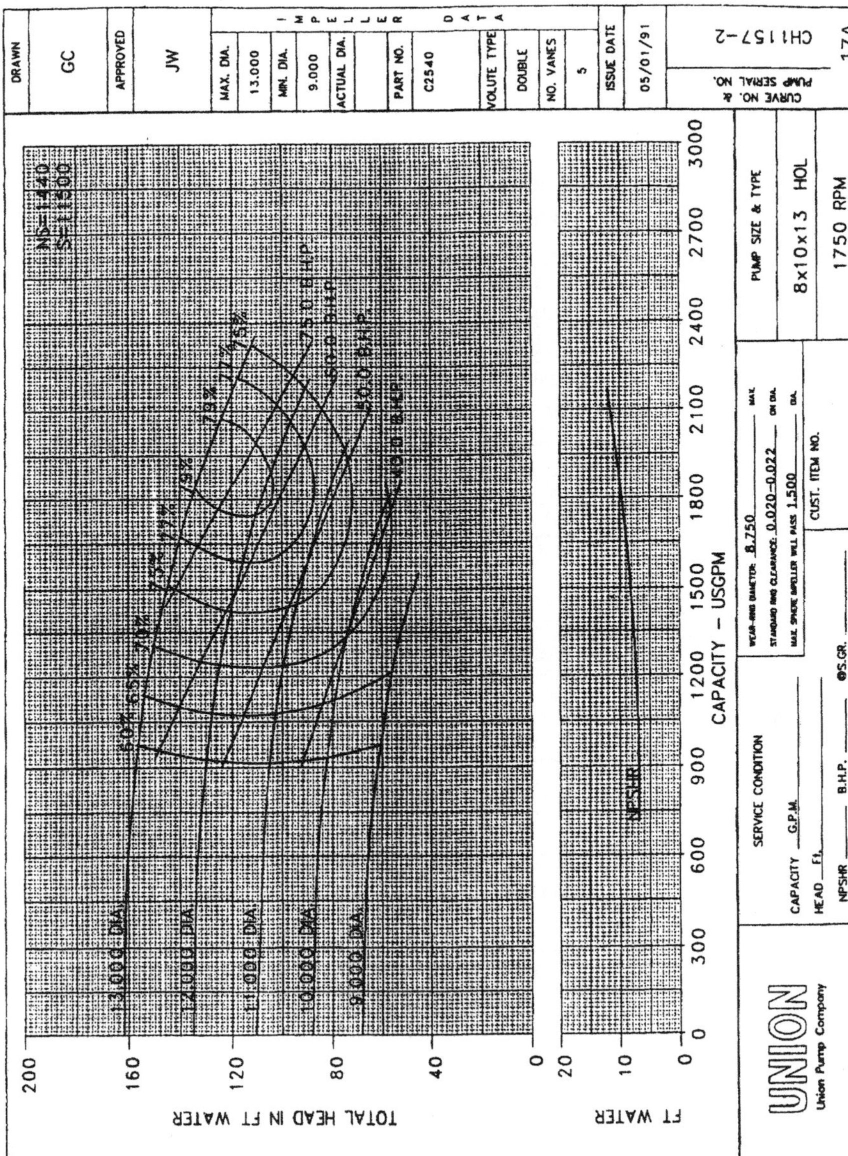

Figure 12.20 Speed change performance verification (Courtesy of Union Pump Company)

The data used for this example is taken from an actual pump operating at 1150 rpm. The affinity law calculation results compare favorably with the same pump operating at 1750 rpm (refer to Figure 12.20).

Protecting the pump

Control and protection are interrelated. Since the objective of the control function is to maximize production throughput at minimum cost, it can sometimes subject the pump to operation in a region which can cause harmful effects to the pump (refer to Figure 12.21).

Consider these factors for pump protection

- Centrifugal pump

  At or below minimum flow operation
  - Overheating
  - High radial loads
  - Internal recirculation damage

  End of curve flow
  - Electric motor overload

  Turbine or engine overspeed
  - Damage to pump/driver

- Positive displacement pump

  Closed discharge valve
  - Damage caused by overpressure
  - Electric motor overload damage

  Overspeed of direct acting pump
  - Damage to pump

Figure 12.21 Consider these factors for pump protection

Pump protection systems

Now that the potential effects of operating a pump outside a "safe region" have been identified, an overview will be provided to illustrate the various methods available to protect each type of pump (refer to Figure 12.22).

Pump protection methods

- Centrifugal pump

  Minimum flow bypass
  - Manual
  - Automatic by external electronic or pneumatic signal

  Motor overload protection
  - Size driver for end of curve power
  - Install orifice in discharge to increase system resistance and limit pump operation
  - Flow or pressure limiting device

  Overspeed (variable speed driver)
  - Governor

- Positive displacement pump

  Overpressure/overload
  - Relief valve sized for full flow

  Overspeed protection (direct acting pump)
  - Governor

Figure 12.22 Pump protection methods

Centrifugal pump protection systems

Since centrifugal pump flow varies inversely to pump head required, they must be protected against low flow operation and high horsepower requirements (high pump flow). Minimum flow bypass systems and proper driver sizing and/or protection systems are therefore required.

Minimum flow bypass protection

Depending on the characteristics of the pumped fluid, continuous operation at low flow conditions can produce centrifugal pump damage. Figure 12.23 shows the three (3) types of minimum flow bypass systems – manual, automatic external and automatic internal. As a rule of thumb, minimum flow protection should be provided for any pump application that can continuously operate below the pump vendor's minimum specified flow rate.

Pump discharge temperature rise calculations should be performed for all multistage, low specific gravity (less than 0.8), and low NPSH available applications to determine if an automatic minimum flow bypass system is required.

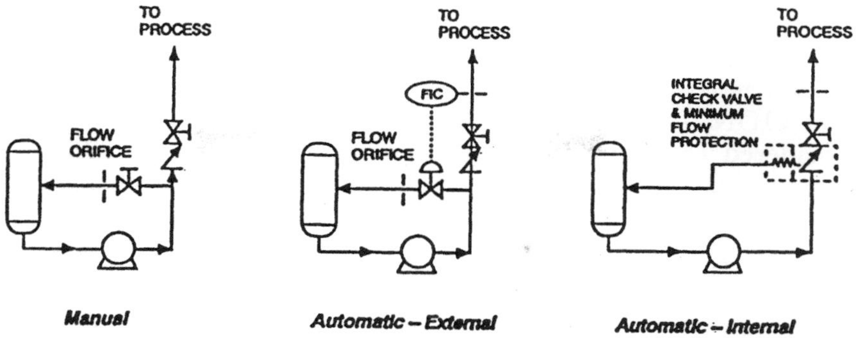

Figure 12.23 Pump minimum flow protection

Automatic minimum flow bypass systems can be either external or internal type but must reliably open at the specified set point. All instruments and valves used must result in repeatable operation. Failure of a minimum bypass system to function when required can cause catastrophic pump damage.

External minimum flow bypass systems incorporate an external sensing transmitter, controller, conventional two way control valve and flow orifice assembly. An internal minimum flow bypass system incorporates all of the above components internal to the minimum flow bypass valve. It is recommended that internal minimum flow bypass valves be used only in clean pumpage systems where the internals will not be affected by corrosion or blockage.

Motor overload protection

If the motor driver is not sized for horsepower capability at the end of the operating curve, the motor will trip on motor overload breaker protection unless some means is provided to prevent an excess horsepower requirement.

One alternative is to intall an orifice to limit flow and therefore power requirements. If an orifice is installed be aware that erosion and/or corrosion is possible and require inspection at turnarounds.

Since horsepower is a function of flow and pressure, these variables can be limited to prevent motor overload.

Automatic motor overload protection can limit motor amps to a preset value by using a controller with input amps to control system level, flow or pressure as required.

Positive displacement pump protection

Since positive displacement pumps are constant flow devices, they must be adequately protected against overpressure. A relief valve usually set to open at 10% above maximum discharge pressure must be provided. **CAUTION – relief valve overpressure (accumulation) must be considered to assure both the pump and the process system are designed for this value.** Overpressure is defined as the pressure a relief valve requires to pass maximum pump flow.

Direct acting positive displacement pumps must be provided with overspeed protection to prevent overspeed in the event of governor malfunction. The liquid end of direct acting positive displacement pumps should also be designed for the maximum stall pressure.

Conversion to metric system

Unit Nomenclature

Symbol	Name	Quantity
a	annum (year)	time
bar	bar	pressure
°C	degree Celsius	temperature
°	degree	plane angle
d	day	time
g	gram	mass
h	hour	time
Hz	hertz	frequency
J	joule	work, energy
K	kelvin	temperature
kg	kilogram	mass
litre	litre	volume
m	metre	length
min.	minute	time
N	newton	force
Pa	pascal	pressure
rad	radian	plane angle
s	second	time
t	tonne	mass
V	volt	electric potential
W	watt	power

Conversion tables – Customary to metric (SI) units

Quantity and SI unit	Customary unit	Metric unit	Conversion factor Multiply customary unit by factor to get metric unit	
Space, time				
Length – m	mi	km	1.609 344	E+00
	m	m	1	
	yd	m	9.144	E–01
	ft	m	3.048	E–01
	in	mm	2.54	E+01
	cm	mm	1.0	E+01
	mm	mm	1	
	mil	µm	2.54	E+01
	micron (µ)	µm	1	
Area – m^2	mi^2	km^2	2.589 988	E+00
	yd^2	m^2	8.361 274	E–01
	ft^2	m^2	9.290 304	E–02
	in^2	mm^2	6.451 6	E+02
	cm^2	mm^2	1.0	E+02
	mm^2	mm^2	1	
Volume, capacity – m^3	m^3	m^3	1	
	yd^3	m^3	7.645 549	E–01
	bbl (42 US gal)	m^3	1.589 873	E–01
	ft^3	m^3	2.831 685	E–02
	UK gal	m^3	4.546 092	E–03
	US gal	m^3	3.785 412	E–03
	litre	dm^3	1	
	UK qt	dm^3	1.136 523	E+00
	US qt	dm^3	9.463 529	E–01
	UK pt	dm^3	5.682 609	E–01
	US pt	dm^3	4.731 765	E–01
	US fl oz	cm^3	2.957 353	E+01
	UK fl oz	cm^3	2.841 305	E+01
	in^3	cm^3	1.638 706	E+01
	ml	cm^3	1	

Quantity and SI unit	Customary unit	Metric unit	Conversion factor Multiply customary unit by factor to get metric unit	
Plane angle – rad	rad	rad	1	
	deg (°)	rad	1.745 329	E–02
	min (′)	rad	2.908 882	E–04
	sec (″)	rad	4.848 137	E–06

Conversion to the metric (SI) system

Quantity and SI unit	Customary unit	Metric unit	Conversion factor Multiply customary unit by factor to get metric unit	
Mass, amount of substance				
Mass – kg	UK ton	t	1.016 047	E+00
	US ton	t	9.071 847	E–01
	UK cwt	kg	5.080 234	E+01
	US cwt	kg	4.535 924	E+01
	kg	kg	1	
	lb	kg	4.535 924	E–01
	oz (troy)	g	3.110 348	E+01
	oz (av)	g	2.834 952	E+01
	g	g	1	
	grain	mg	6.479 891	E+01
	mg	mg	1	
	μg	μg	1	
Amount of substance – mol	lb mol	kmol	4.535 924	E–01
	g mol	kmol	1.0	E–03
	std m^3 (0°C, 1atm)	kmol	4.461 58	E–02
	std m^3 (15°C, 1atm)	kmol	4.229 32	E–02
	std ft^3 (60°C, 1 atm)	kmol	1.195 30	E–03

Quantity and SI unit	Customary unit	Metric unit	Conversion factor Multiply customary unit by factor to get metric unit	
Calorific value, heat, entropy, heat capacity				
Calorific value – J/kg	Btu/lb	MJ/kg	2.326 000	E–03
(Mass basis)		kJ/kg	2.326 000	E+00
	cal/g	kJ/kg	4.184	E+00
	cal/lb	J/kg	9.224 141	E+00
Calorific Value –J/mol	kcal/g mol	kJ/kmol	4.184	E+03
(Mole basis)	Btu/lb mol	MJ/kmol	2.326 000	E–03
		kJ/kmol	2.326 000	E+00
Calorific value – J/m^3	therm/UK gal	MJ/m^3	2.320 800	E+04
(Volume basis –		kJ/m^3	2.320 800	E+07
solids and liquids)	Btu/US gal	MJ/m^3	2.787 163	E–01
		kJ/m^3	2.787 163	E+02
	Btu/UK gal	MJ/m^3	2.320 800	E–01
		kJ/m^3	2.320 800	E+02
	Btu/ft^3	MJ/m^3	3.725 895	E–02
		kJ/m^3	3.725 895	E+01
	kcal/m^3	MJ/m^3	4.184	E–03
		kJ/m^3	4.184	E+00
	cal/ml	MJ/m^3	4.184	E+00
	ft•lb$_f$/US gal	kJ/m^3	3.581 692	E–01
Calorific value – J/m^3	cal/ml	kJ/m^3	4.184	E+03
(Volume basis – gases)	kcal/m^3	kJ/m^3	4.184	E+00
	Btu/ft^3	kJ/m^3	3.725 895	E+01
Specific entropy – J/kg•K	Btu/lb•°R	kJ/kg•K	4.186 8	E+00
	cal/g•°K	kJ/kg•K	4.184	E+00
	kcal/kg•°C	kJ/kg•K	4.184	E+00
Specific heat	kW·h/kg•°C	kJ/kg•°C	3.6	E+03
capacity – J/kg•K	Btu/lb•°F	kJ/kg•°C	4.186 8	E+00
(Mass basis)	kcal/kg•°C	kJ/kg•°C	4.184	E+00
Specific heat	Btu/lb mol•°F	kJ/kmol•°C	4.186 8	E+00
capacity – J/mol•K	cal/g mol•°C	kJ/kmol•°C	4.184	E+00
(Mole basis)				

Quantity and SI unit	Customary unit	Metric unit	Conversion factor Multiply customary unit by factor to get metric unit	
Temperature, pressure, vacuum				
Temperature – K	°R	K	5/9	
(Absolute)	°K	K	1	
Temperature – K	°F	°C	5/9 (°F – 32)	
(Traditional)	°C	°C	1	
Temperature – K	°F	°C	5/9	
(Difference)	°C	°C	1	
Temperature/Length – K/m (Geothermal gradient)	°F per 100 ft	mK/m	1.822 689	E+01
Pressure – Pa	atm	MPa	1.013 250	E–01
		kPa	1.013 250	E+02
	bar	MPa	1.0	E–01
		kPa	1.0	E+02
	at (kg_f/cm^2)	MPa	9.806 650	E–02
		kPa	9.806 650	E+01
	lb_f/in^2 (psi)	MPa	6.894 757	E–03
		kPa	6.894 757	E+00
	in Hg (60°F)	kPa	3.376 85	E+00
	in H_2O (39.2°F)	kPa	2.490 82	E–01
	in H_2O (60°F)	kPa	2.488 4	E–01
	mm Hg = torr (0°C)	kPa	1.333 224	E–01
	cm H_2O (4°C)	kPa	9.806 38	E–02
	lb_f/ft^2 (psf)	kPa	4.788 026	E–02
	μm Hg (0°C)	Pa	1.333 224	E–01
	μbar	Pa	1.0	E–01
	dyn/cm^2	Pa	1.0	E–01
Vacuum, draft – Pa	in Hg (60°F)	kPa	3.376 85	E+00
	in H_2O (39.2°F)	kPa	2.490 82	E–01
	in H_2O (60°F)	kPa	2.488 4	E–01
	mm Hg = torr (0°C)	kPa	1.33 224	E–01
	cm H_2O (4°C)	kPa	9.806 38	E–02
Liquid head – m	ft	m	3.048	E–01
	in	mm	2.54	E+01
Pressure drop/length – Pa/m	psi/ft	kPa/m	2.262 059	E+01
	psi/100 ft	kPa/m	2.262 059	E–01

Quantity and SI unit	Customary unit	Metric unit	Conversion factor Multiply customary unit by factor to get metric unit	
Density, specific volume, concentration, dosage				
Density (gases) – kg/m^3	lb/ft^3	kg/m^3	1.601 846	E+01
		g/m^3	1.601 846	E+04
Density (liquids) – kg/m^3	lb/US gal	kg/dm^3	1.198 264	E–01
	lb/UK gal	kg/dm^3	9.977 644	E–02
	lb/ft^3	kg/dm^3	1.601 846	E–02
	g/cm^3	kg/dm^3	1	
Density (solids) – kg/m^3	lb/ft^3	kg/dm^3	1.601 846	E–02
Specific volume – m^3/kg (gases)	ft^3/lb	m^3/kg	6.242 796	E–02
		m^3/g	6.242 796	E–05
Specific volume – m^3/kg (liquids)	ft^3/lb	dm^3/kg	6.242 796	E+01
	UK gal/lb	dm^3/kg	1.022 241	E+01
	US gal/lb	dm^3/kg	8.345 404	E+00
Specific volume – m^3/mol (mole basis)	litre/g mol	$m^3/kmol$	1	
	ft^3/lb mol	$m^3/kmol$	6.242 796	E–02
Concentration – kg/kg (mass/mass)	wt %	kg/kg	1.0	E–02
		g/kg	1.0	E–05
	wt ppm	mg/kg	1	
Concentration – kg/m^3 (mass/volume)	lb/bbl	kg/m^3	2.853 010	E+00
	g/US gal	kg/m^3	2.641 720	E–01
	g/UK gal	kg/m^3	2.199 692	E–01

Quantity and SI unit	Customary unit	Metric unit	Conversion factor Multiply customary unit by factor to get metric unit	
Facility throughput, capacity				
Throughput – kg/s	million lb/yr	t/a	4.535 924	E+02
(mass basis)	UK ton/yr	t/a	1.016 047	E+00
	US ton/yr	t/a	9.071 847	E-01
	UK ton/d	kg/h	1.016 047	E+00
	US ton/d	t/d	9.071 847	E-01
	UK ton/h	t/h	1.016 047	E+00
	US ton/h	t/h	9.071 847	E-01
	lb/h	t/d	4.535 924	E-01
Throughput – m^3/s	bbl/d	t/a	5.803 036	E+01
(volume basis)		m^3/h	6.624 471	E-03
	ft^3/d	m^3/h	1.179 869	E-03
	bbl/h	m^3/h	1.589 873	E-01
	ft^3/h	m^3/h	2.831 685	E-02
	UK gal/h	m^3/h	4.546 092	E-03
	US gal/h	m^3/h	3.785 412	E-03
	UK gal/min	m^3/h	2.727 655	E-01
	US gal/min	m^3/h	2.271 247	E-01
Throughput – mol/s (mole basis)	lb mol/h	kmol/h	4.535 924	E-01

Quantity and SI unit	Customary unit	Metric unit	Conversion factor Multiply customary unit by factor to get metric unit	
Flow rate				
Flow rate – kg/s (mass basis)	UK ton/min	kg/s	1.693 412	E+01
	US ton/min	kg/s	1.511 974	E+01
	UK ton/h	kg/s	2.822 353	E–01
	US ton/h	kg/s	2.519 958	E–01
	UK ton/d	kg/s	1.175 980	E–02
	US ton/d	kg/s	1.049 982	E–02
	million lb/yr	kg/s	5.249 912	E+00
	UK ton/yr	kg/s	3.221 864	E–05
	US ton/yr	kg/s	2.876 664	E–05
	lb/s	kg/s	4.535 924	E–01
	lb/min	kg/s	7.559 873	E–03
	lb/h	kg/s	1.259 979	E–04
Flow rate – m^3/s (volume basis)	bbl/d	dm^3/s	1.840 131	E–03
	ft^3/d	dm^3/s	3.277 413	E–04
	bbl/h	dm^3/s	4.416 314	E–02
	ft^3/h	dm^3/s	7.865 791	E–03
	UK gal/h	dm^3/s	1.262 803	E–03
	US gal/h	dm^3/s	1.051 503	E–03
	UK gal/min	dm^3/s	7.576 820	E–02
	US gal/min	dm^3/s	6.309 020	E–02
	ft^3/min	dm^3/s	4.719 474	E–01
	ft^3/s	dm^3/s	2.831 685	E+01
Flow rate – mol/s (mole basis)	lb mol/s	kmol/s	4.535 924	E–01
	lb mol/h	kmol/s	1.259 979	E–04

Quantity and SI unit	Customary unit	Metric unit	Conversion factor Multiply customary unit by factor to get metric unit	
Energy, work, power				
Energy, work – J	therm	MJ	1.055 056	E+02
		kJ	1.055 056	E+05
		kJ	3.6	E+03
	Btu	kJ	1.055 056	E+00
	kcal	kJ	4.184	E+00
	cal	kJ	4.184	E–03
	$ft \bullet lb_f$	kJ	1.355 818	E–03
	lb•ft	kJ	1.355 818	E–03
	J	kJ	1.0	E–03
	$lb \bullet ft^2/s^2$	kJ	4.214 011	E–05
	erg	J	1.0	E–07
Power – W	million Btu/h	MW	2.930 711	E–01
	ton of refrigeration	kW	3.516 853	E+00
	Btu/s	kW	1.055 056	E+00
	kW	kW	1	
	hydraulic horsepower–hhp	kW	7.460 43	E–01
	hp (electric)	kW	7.46	E–01
	Btu/min	kW	1.758 427	E–02
	$ft \bullet lb_f/s$	kW	1.355 818	E–03
	kcal/h	W	1.162 222	E+00
	Btu/h	W	2.930 711	E–01
	$ft \bullet lb_f/min$	W	2.259 697	E–02
Power/area – W/m^2	$Btu/s \bullet ft^2$	kW/m^2	1.135 653	E+01
	$Cal/h \bullet cm^2$	kW/m^2	1.162 222	E–02
	$Btu/h \bullet ft^2$	kW/m^2	3.154 591	E–03
Cooling duty – W/W (machinery)	Btu/bhp•h	W/kW	3.930 148	E–01
Specific fuel – kg/J consumption (mass basis)	lb/hp•h	mg/J	1.689 659	E–01
Specific fuel – m^3/J consumption (volume basis)	$m^3/kW \bullet h$	dm^3/MJ	2.777 778	E+02
	US gal/hp•h	dm^3/MJ	1.410 089	E+00

Quantity and SI unit	Customary unit	Metric unit	Conversion factor Multiply customary unit by factor to get metric unit	
Mechanics				
Velocity (linear) – m/s	mi/h	km/h	1.609 344	E+00
speed	m/s	m/s	1	
	ft/s	m/s	3.048	E–01
	ft/min	m/s	5.08	E–03
	ft/h	mm/s	8.466 667	E–02
	in/s	mm/s	2.54	E+01
	in/min	mm/s	4.233 333	E–01
Corrosion rate – mm/a	in/yr (ipy)	mm/a	2.54	E+01
Rotational frequency – rev/s	rev/s	rev/s	1	
	rev/min	rev/s	1.666 667	E–02
Acceleration – m/s^2 (linear)	ft/s^2	m/s^2	3.048	E–01
	gal (cm/s^2)	m/s^2	1.0	E–02
Acceleration – rad/s^2 (rotational)	rad/s^2	rad/s^2	1	
Momentum – kg•m/s	lb•ft/s	kg•m/s	1.382 550	E–01
Force – N	UK ton$_f$	kN	9.964 016	E+00
	US ton$_f$	kN	8.896 443	E+00
	kg$_f$ (kp)	N	9.806 650	E+00
	lb$_f$	N	4.448 222	E+00
	N	N	1	
Bending moment, – N•m torque	US ton$_f$•ft	kN•m	2.711 636	E+00
	kg$_f$•m	N•m	9.806 650	E+00
	lb$_f$•ft	N•m	1.355 818	E+00
	lb$_f$•in	N•m	1.129 848	E–01
Bending – N•m/m moment/length	lb$_f$•ft/in	N•m/m	5.337 866	E+01
	kg$_f$•m/m	N•m/m	9.806 650	E+00
	lb$_f$•in/in	N•m/m	4.448 222	E+00
Moment of – kg•m^2 inertia	lb$_f$•ft^2	kg•m^2	4.214 011	E–02
	in^4	cm^4	4.162 314	E+01

Quantity and SI unit	Customary unit	Metric unit	Conversion factor Multiply customary unit by factor to get metric unit	
Mechanics continued				
Stress – Pa	US ton$_f$/in^2	MPa	1.378 951	E+01
	kg$_f$/mm^2	MPa	9.806 650	E+00
	US ton$_f$/ft^2	MPa	9.576 052	E–02
	lb$_f$/in^2 (psi)	MPa	6.894 757	E–03
	lb$_f$/ft^2 (psf)	kPa	4.788 026	E–02
	dyn/cm^2	Pa	1.0	E–01
Mass/length – kg/m	lb/ft	kg/m	1.488 164	E+00
Transport properties				
Thermal resistance – K•m^2/W	°C•m^2•h/kcal	°C•m^2/kW	8.604 208	E+02
	°F•ft^2•h/Btu	°C•m^2/kW	1.761 102	E+02
Heat flux – W/m^2	Btu/h•ft^2	kW/m^2	3.154 591	E–03
Thermal – W/m•K conductivity	cal/s•cm^2•°C/cm	W/m•°C	4.184	E+02
	Btu/h•ft^2•°F/ft	W/m•°C	1.730 735	E+00
	Kcal/h•m^2•°C/m	W/m•°C	1.162 222	E+00
	Btu/h•ft^2•°F/in	W/m•°C	1.442 279	E–01
	cal/h•cm^2•°C/cm	W/m•°C	1.162 222	E–01
Heat transfer – W/m^2•K coefficient	cal/s•cm^2•°C	kW/m^2•°C	4.184	E+01
	Btu/s•ft^2•°F	kW/m^2•°C	2.044 175	E+01
	cal/h•cm^2•°C	kW/m^2•°C	1.162 222	E–02
	Btu/h•ft^2•°F	kW/m^2•°C	5.678 263	E–03
	Btu/h•ft^2•°R	kW/m^2•K	5.678 263	E–03
	kcal/h•m^2•°C	kW/m^2•°C	1.162 222	E–03
Volumetric heat – W/m^3•K transfer coefficient	Btu/s•ft^3•°F	kW/m^3•°C	6.706 611	E+01
	Btu/h•ft^3•°F	kW/m^3•°C	1.862 947	E–02
Surface tension – N/m	dyn/cm	mN/m	1	
Viscosity – Pa•s (dynamic)	lb$_f$•s/in^2	Pa•s	6.894 757	E+03
	lb$_f$•s/ft^2	Pa•s	4.788 026	E+01
	kg$_f$•s/m^2	Pa•s	9.806 650	E+00
	lb/ft•s	Pa•s	1.488 164	E+00
	dyn•s/cm^2	Pa•s	1.0	E+01
	cP	Pa•s	1.0	E–03
	lb/ft•h	Pa•s	4.133 789	E–04

Quantity and SI unit	Customary unit	Metric unit	Conversion factor Multiply customary unit by factor to get metric unit	
Transport properties continued				
Viscosity – m^2/s	ft^2/s	mm^2/s	9.290 304	E+04
(kinematic)	in^2/s	mm^2/s	6.451 6	E+02
	m^2/h	mm^2/s	2.777 778	E+02
	cm^2/s	mm^2/s	1.0	E+02
	ft^2/h	mm^2/s	2.580 64	E+01
	cSt	mm^2/s	1	

Index

FSC
www.fsc.org
MIX
Papier | Fördert gute Waldnutzung
FSC® C083411

Kolibri 360 GmbH
Ferdinand-Jühlke-Straße 7
99095 Erfurt, Deutschland
produktsicherheit@kolibri360.de